CUSTOM AUTO WIRING & ELECTRICAL

**OEM Electrical Systems, Premade & Custom Wiring Kits, & Car Audio
Installations for Street Rods, Muscle Cars, Race Cars, Trucks & Restorations**

Matt Strong

HPBOOKS

HPBooks

Published by the Penguin Group
Penguin Group (USA) Inc.
375 Hudson Street, New York, New York 10014, USA

Penguin Group (Canada), 90 Eglinton Avenue East, Suite 700, Toronto, Ontario M4P 2Y3, Canada
(a division of Pearson Penguin Canada Inc.)
Penguin Books Ltd., 80 Strand, London WC2R 0RL, England
Penguin Group Ireland, 25 St. Stephen's Green, Dublin 2, Ireland (a division of Penguin Books Ltd.)
Penguin Group (Australia), 250 Camberwell Road, Camberwell, Victoria 3124, Australia
(a division of Pearson Australia Group Pty. Ltd.)
Penguin Books India Pvt. Ltd., 11 Community Centre, Panchsheel Park, New Delhi—110 017, India
Penguin Group (NZ), 67 Apollo Drive, Rosedale, North Shore 0632, New Zealand
(a division of Pearson New Zealand Ltd.)
Penguin Books (South Africa) (Pty.) Ltd., 24 Sturdee Avenue, Rosebank, Johannesburg 2196, South Africa

Penguin Books Ltd., Registered Offices: 80 Strand, London WC2R 0RL, England

While the author has made every effort to provide accurate telephone numbers and Internet addresses at the time of publication, neither the publisher nor the author assumes any responsibility for errors, or for changes that occur after publication. Further, the publisher does not have any control over and does not assume any responsibility for author or third-party websites or their content.

CUSTOM AUTO WIRING & ELECTRICAL

ISBN: 978-1-55788-545-6

PRINTED IN THE UNITED STATES OF AMERICA

10   9   8   7   6   5   4   3   2

NOTICE: The information in this book is true and complete to the best of our knowledge. All recommendations on parts and procedures are made without any guarantees on the part of the author or the publisher. Tampering with, altering, modifying or removing any emissions-control device is a violation of federal law. Author and publisher disclaim all liability incurred in connection with the use of this information. We recognize that some words, engine names, model names and designations mentioned in this book are the property of the trademark holder and are used for identification purposes only. This is not an official publication.

CONTENTS

ACKNOWLEDGMENTS

When I received a call from HPBooks to write this book, I was ecstatic. At the time, I thought it would really be as simple as writing a bunch of magazine articles with a common thread tying them together. I had no idea that I would have to learn to format my writing in a very specific manner. My editor, Michael Lutfy, not only gave me the opportunity, but he also patiently guided me through the unique process of writing a technical book, helping to shape what had become a collection of independent articles into a cohesive, structured manuscript on a difficult subject. He graciously extended deadline after deadline as I struggled to revise and reorganize the material and photos, a process greatly hampered by my limited mobility after shoulder surgery. I thank him for his faith in me.

I also had tremendous help from my brother-in-law, Ian Mowat, who has worked in electronics and with electrical circuitry all his professional life. He is the brother I've always wanted and the task of writing this book would have been much more difficult without him. From helping me to create diagrams, to proofreading and prodding me to get it done, his support has been invaluable. His expertise has helped me explain how a bunch of invisible electrons move down a conductor, how the process works and how it can go wrong, in layman's terms.

I also have to recognize the assistance from COMP Cams, Fuel Air Spark Technology (F.A.S.T.), FlowTech Headers, Holley, M&H Electric Fabricators, MSD Ignition, Painless Performance, RW Industries, Snap-on Tools, Stewart Warner and YearOne. Each of these companies contributed their technology, knowledge and product(s). Without their help this book would not have been possible.

Finally, my wife and two gearhead sons have been the anchor I've clung to in the rough weather I've had to overcome while writing this book. They always asked how things were going, were always very supportive, even when I was in denial about the RIGHT way to write a book.

INTRODUCTION

Writing DIY auto books is a difficult task, since it must present a subject to a diverse group of gearheads, teach them as much as possible of the theory without losing their interest, then illustrate how to do it step by step in terms general enough to be applied to any automotive project. It also must be timely and fun to read. For example, in today's world it is not important to understand how to rebuild a generator. We use alternators and when they wear out they get replaced with a new or remanufactured unit. Therefore, that old generator subject isn't needed anymore.

Wiring a car from scratch is a daunting task for nearly everyone but the most experienced electrical expert. Overall, the task seems complex, but when you break down electrical systems into smaller bits, you'll find it isn't all that difficult. I had the opportunity to do three complex projects for this book, and I found the process to be very rewarding, an experience I hope you'll have with your own project.

If you purchased this book, then it means you are trying to increase your knowledge and understanding of the electrons running through the maze of wires in your vehicle. Of course, if you are starting from scratch you want to know where and why wires should go in a particular location and how to complete circuits. This book was written as a modern reference book about automotive electrical systems. With the information inside, I'm confident you can learn to wire a complete vehicle from scratch, start to finish, or just fix a circuit that isn't behaving properly. It doesn't matter if it is in a hot rod, street rod, muscle car, race car or family sedan, this book will show you how to wire almost any vehicle.

I am assuming that like most readers of HPBooks, you will already have the basic tools needed to work on cars. Any specilized electrical tools, like mulitimeters, volt meters, strippers and crimpers, are covered within the context they are used. The first chapters cover the basics of electricity and define terms like amperes, volts and watts and take you into reading the road maps of electrical circuits and wiring diagrams. They are the foundation you need to build your confidence and knowledge base to a point that you are ready to jump right into a project. From here we jump right into big electrical projects with step-by-step photos and text to walk you through almost any electrical installation you may have in mind for your own vehicle. We wire a Pro Street 1956 Ford pickup from start to finish. We replace every single wire in a 1971 Barracuda, we integrate a new wiring kit into the stock engine management system on an Acura race car and finally, we install two electronic fuel injection systems into two different vehicles. We also do a high-end radio installation for a modern vehicle, plus review the radio installation in the 1956 Ford pickup. We wrap things up with a section on troubleshooting tips that can be used in any kind of vehicle, from grocery getter to race car.

Remember that the car hobby is supposed to be fun, and that even goes for electrical projects. Your investment in time and energy will make you a better mechanic and hot rodder. Perhaps the greatest reward you can receive is the satisfaction from having done the job yourself. Hopefully, this book will help you get there.

—Matt Strong

Chapter 1

Basic Auto Electrical Theory

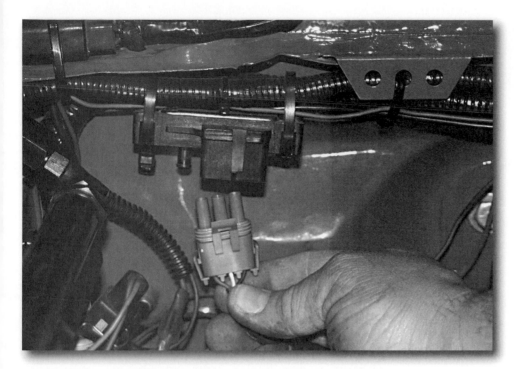

When you are wiring up an EFI system, such as connecting the MAP sensor as shown here, you need to understand basic electricity to know why you do things as well as what to do. Instructions from suppliers are good enough to get you through the installation if you are patient, but understanding what things do will allow you to do a great job and have everything looking as good as it runs.

Electrical theory can be very boring and confusing to some folks, mainly because you cannot see the electrons running through wires, relays, coils or other elements of a vehicle electrical system. For that reason we are not going deep into theory, but we are giving you more than enough to understand electricity in any circuit; car, truck or even your home. In this chapter we'll cover the principles of voltage, resistance and circuits. Later, when you are working on your own project, they will help you focus upon important details and you will understand why you need to be doing specific things. We are also using examples of fluid circuits to explain how electricity does its job. Take your time reading this chapter and you will find it contains a lot of useful information.

Understanding Electrical Flow

Electricity behaves a lot like a fluid; we say it *flows* through the wires. It runs from one battery terminal to the other and all the wires between them when called upon, just like the fluid in a hydraulic (liquid) system. The fluid pump is the battery or alternator and the fluid is under pressure. How much pressure depends on the pump's power. The fluid delivers energy to components and makes them work. If you have a large hydraulic ram, you need a large pipe to feed in the power. But for smaller rams a smaller pipe will be fine. To both, though, the same pressure is delivered (usually between 1,000 and 3,000 psi).

The higher the pressure, the smaller the ram can be so the equipment can be a little lighter and the fluid volume will be less, but the hoses are still sized to the flow, just like electrical systems must be. It is much the same for electricity; 24 volts can deliver more power than 12 volts, but automotive vehicles are all 12 volts to save weight and cost. The pressure (voltage) in any given circuit is the same no matter what the pipe (wire) size and it delivers energy to the components such as the starter motor so they can do their work. But if you are delivering 1,000 volts through a large wire with high amperage, it can do far more work than 12 volts through a 16-gauge wire.

It can also do far more damage if the circuits are not designed to handle its potential. Liquid flow consists of two components—pressure and volume. With electricity, the "pressure" is voltage and "volume" is amperage. The battery provides the voltage (pressure) and the wires the amperage (volume). The more electricity you need for a component, the fatter the wire you need to feed it. If a pipe is not big enough it will explode. If a wire is not big enough it will also fail, generally by burning up.

You will find bundles of wires, or you will create them, as you go through the process of fixing or modifying your project vehicle. It is actually easier to start with a vehicle that doesn't have a single wire in it, than to work around an existing electrical system. When we did the how-to sections for this book, one of the vehicles was just a complete chassis

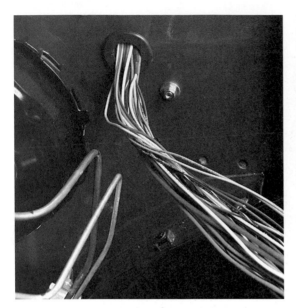

This is the main wiring harness running through a vehicle firewall out into the engine compartment. Since it is a pickup truck, this is also where the wires start as they run to the back of the vehicle. These are just like the many pipes that run through your home and throughout your community to carry fluids. They look complicated in a big bunch, but when you take the time to read what each does they are simple to follow.

These are long battery cables for mounting the battery in the rear of a vehicle. They weigh close to 16 lbs. if used at their full length. By grounding the battery at the rear on the frame, the pickup was able to save about 8 lbs. in cable weight.

All four wires in the top photo will deliver 12 volts, but the large wire on top can do far more work than the small one because it can carry more amperage. In the bottom photo a 10-gauge wire was placed on top of a battery cable to show just how much bigger a battery cable must be to handle the requirements of starting a vehicle and keeping it running. At start-up, especially in cold weather, the system can draw as much as 75 to 90 amperes.

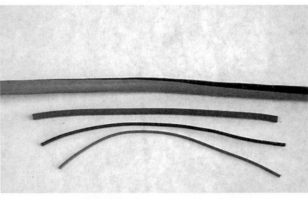

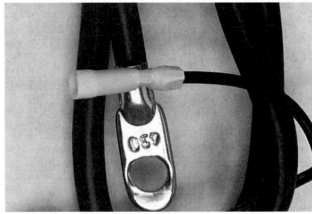

and an empty truck cab. It was a snap. But, when we had to integrate new wiring into the stock system in an Acura hatchback, the job really got tough.

Electrical Terms

Current—The number of electrons is called current and its unit of measurement is the ampere or amp for short. Electrical current is like the rate that water flows through a pipe. If there is a large

volume of water to flow, you need a big pipe; a small amount of water can make do with a much smaller size pipe. I prefer to think of electrons as little people trying to get through an airport. There are millions of them and they do fine as long as the flow isn't bottlenecked as it is for security. That is how resistance builds—with bottlenecks in the circuit.

Resistance—An ohm is the unit of measurement of electrical resistance. A conductor, like a piece of metal, has its atoms arranged so electrons can readily pass around the atoms with little friction or resistance. In a non-conductor or poor conductor, the atoms are arranged to greatly resist or impede the travel of the electrons. This resistance is similar to the friction of a hose against the water moving through it. But in the case of non-conductors, electricity will not pass through the material; it is just like a pipe cap or shut-off valve that stops flow cold. But resistance, even through a good conductor, can get too high if the wire is too long (front to rear battery changes) or it is corroded inside. (The battery cables, right at the battery, can corrode in the copper wires and you can see it as the wire turns green.) If the cable is well sealed near the battery, where gasses vent, this will be less of a problem.

Grounds—Ground is the term used for the electrical return in any automotive circuit. The term comes from the early use of lightning rods from the highest part of a building running down into the earth, pioneered by Benjamin Franklin. (In England, a ground is called "earth" still.) Even today many homes rely on a rod pushed into the ground in the base of the house for its earth rail.

In home electrics it is different because AC

The stud being installed on the floor of this race car is a common ground point for all circuits inside the vehicle that need a one. A larger wire is run from the stud to ground at the battery negative post. If something stops working, it is easy to check the grounds.

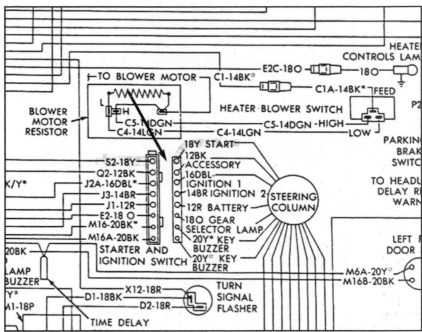

The two long rectangles to the left of the round steering column (arrow) on this wiring diagram have the same wire numbers in exactly the same position from top to bottom. This tells you this is the steering column main connector. It also shows you the colors and wire size for each. This is just a small part of what a circuit diagram delivers.

circuits need two wires to conduct electricity while cars use DC, which is always running from positive to negative. The negative is called the ground because it does not generally have a separate wire; it uses the chassis and engine to complete the circuit back to the battery.

Consider this: there is only one wire for each spark plug. The positive lead is from the plug wire. After the current jumps to make a spark, the return of the circuit runs through the engine block, the engine ground strap, chassis and finally back to the negative terminal on the battery.

This method allows engineers to use the whole of the chassis, engine and body as a return circuit—assuming the car has a metal body. This also minimizes wire requirements and weight. Plastic body cars need wires from components to a good ground. But they are usually bundled to a common ground to minimize wire usage and weight.

Circuits—Think of circuits as pipes to guide electrons to where they are needed. If you look at a wiring diagram, you will see a lot of little lines running from place to place and labels like BR or PK, 10A or 20A all over the diagram. There should also be names like headlight, cigarette lighter, etc. While wiring diagrams and their notations can be confusing, if you look at most diagrams you will find that the "BR" is for Brown, and "PK" is for pink, in other words they are codes for wire colors. Some might even have a stripe down them (called a tracer color). The "10A" or "20A" notations are for the size fuse or circuit breaker the circuit requires. It is not so hard once you think it through.

Wire Sizes—Because cars run on just 12 volts, the current required for things like headlamps is large. When the current is high—10s of amps—it is necessary to have thicker wires. Thin wires will act like the filament in a heater and get so hot they will eventually burst into flames if they are asked to carry too large a load. Never be afraid to open a harness if you are having problems identifying a circuit or wire. A wiring diagram will tell you what to look for and the position in the vehicle where it is located. Just open it up (without doing any damage to the wires inside) and have a look. You need to make sure you know exactly how much current is running through a circuit and select the correct fuse and wire size for it, plus a little margin for error. The calculation is simple.

$I = W \div V$ (I is amps, W is watts and V volts)

In other words, if you have a headlamp at 48 watts the current it needs is $48 \div 12 = 4$ amps. But there are normally two headlamps in the same circuit. Therefore the current is $4 + 4 = 8$ amps and you need to select a 10-amp fuse. But, if the same circuit also powers the parking and running lights you might need a larger fuse. Probably around 15A to 20A in most cases, but do the math and be sure. Too large a fuse might allow too much current to flow through small wires and cause damage.

Wire Gauges—Wire sizes are listed in terms of

AMPS	WIRE LENGTH						
	3'	5'	7'	10'	15'	20'	25'
5 Amps	18	18	18	18	18	18	18
6	18	18	18	18	18	18	18
7	18	18	18	18	18	18	18
8	18	18	18	18	18	16	16
10	18	18	18	18	16	16	16
11	18	18	18	18	16	16	14
12	18	18	18	18	16	16	14
15	18	18	18	18	14	14	12
18	18	18	16	16	14	14	12
20	18	18	16	16	14	12	10
22	18	18	16	16	12	12	10
24	18	18	16	16	12	12	10
30	18	16	16	14	10	10	10
40	18	16	14	12	10	10	8
50	16	14	12	12	10	10	8
100	12	12	10	10	6	6	4
150	10	10	8	8	4	4	2
200	10	8	8	6	4	4	2

This table tells you the size (gauge) wire you need for the given amps and distance or length of the wire. The gauges are the shaded numbers.

This harness is made from a Painless Performance kit and all their wires have their function written right on the wire. This makes identification easy. In an OEM harness you must have a wiring diagram to identify the correct color for any wire inside.

This side-connect GM battery actually delivers up to 540 cranking amps to the starter at 32°F and 450 amps at below zero temperatures. That amount of amperes should blow out every circuit in the vehicle, but that is a flash amperes rating and it is for only the smallest fraction of a second as it hits the starter solenoid to turn the starter motor.

gauge—the smaller the number, the thicker the wire. Please study the wire gauge table above to guide you on wire gauge selection. Note that you will need larger wires when the length is extended. The longer the run, the greater the resistance, which leads to a voltage drop. To make up for that you can go up one size such as from an 18 gauge (ga.) to a 16 gauge. Generally you will find wire gauges from 18ga. (small) to 10ga. (large) throughout the vehicle. But battery cables will be much bigger with size 4ga. to 2ga. or larger doing the work to and from the battery. Just remember that the more load (amperage) a wire needs to carry, the larger it must be.

12-Volt Automotive Components

Note: Some of the following components will be covered in greater detail in the Chapter 2, but they are presented here as an overview of a basic 12-volt automotive electrical system. The electricity in a car is 12-volt direct current (DC), which simply means it is always flowing in the same direction and at the same voltage or pressure. It flows from positive to negative, from plus (+) to minus (–), or red to black. Remember this because it is the basis

for all electrical work on a car.

Potential difference = voltage = a slope with water stored at the top

Batteries—Batteries come in several styles and with different technology. They may be traditional, with flat plates and separators, or have round plate technology as seen in the Optima line. Regardless of their shape and style, they all do the same thing—store and deliver electricity in an on-demand basis.Batteries are covered in greater detail on page 6.

Alternators—The alternator supplies the electricity in modern cars, using fields and magnets

Alternators come in different shapes, but their function is the same—making AC current that is then sent through a rectifier to convert it to DC current for the vehicle. (These rectifiers are an integral part of the alternator in all but the oldest models.) The other alternator shot is a custom unit designed for high-amperage systems. Today's electronics place quite a load on the vehicle, so high-output alternators are needed. High output, or high amperage refers to the alternator's ability to deliver more amps/power to the vehicle when the demand is high.

The starter motor will draw a lot of amperage (working force) from the battery during starting, especially in cold weather, so you have to have a large enough alternator to make sure it gets the power it needs. This is a high-torque starter for engines that have high compression for racing or operate in a working environment that is cold and harsh.

to generate electrical current from circular motion. It creates AC and needs a rectifier/converter to change the current to DC. Alternators are used because they are more efficient, especially at low engine speeds. Older vehicles use a generator, which creates DC current directly. We will not cover these or other old-school parts, as they have few applications in today's automotive world, except for the ground-up original restoration of classic vehicles.

Some specialty vehicles may have a second alternator mounted on the front of the engine powered by a drive belt. Some even drive the belt by connecting it to the driveshaft under the car. Keep in mind that this type of installation will not generate electricity unless the vehicle is moving at a decent speed.

Starter Motor—The starter motor is the largest most powerful electric component on a car. We will not go into how to rebuild a starter motor or generator because the parts are not readily available, plus rebuilt starters and alternators are cheap and come with a guarantee. But remember the starter is one of the most critical parts when it comes to getting a vehicle started. By the way, an electric engine delivers maximum torque the instant it gets electricity from the system. That is why electric automotive motors can power a vehicle from 0–30 mph so quickly.

Coils—Coils in a vehicle increase voltage from 12 volts to around +25,000 volts to make electricity jump from one place to another, such as from the spark plug ceramic/wire to the ground electrode on the metal part of the plug. Some high-performance coils can deliver +40,000 volts to fire the plugs of a high-compression modified engine. These high-performance coils take 12 volts and convert it to +40,000 volts by having more field windings before the current is sent to the spark plugs, therefore they cost more. They also generate a lot of heat in the conversion process so they have oil inside to cool the coil or another strategy to do the cooling.

Battery Technology—A battery consists of a sealed box of 6 separate cells. Each has two plates and an electrolyte, usually a strong solution of acid. Each cell produces nearly 2 volts and when placed

The coil on the left looks like a chrome standard coil, but it is actually capable of delivering +40,000 volts to a spark plug. The coil on the right looks like what it really is; a powerful high-performance coil also delivering +40,000 volts to a spark plug.

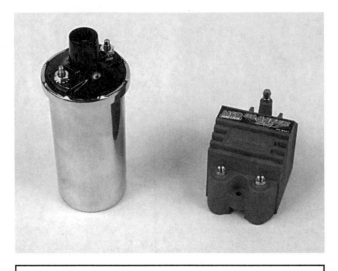

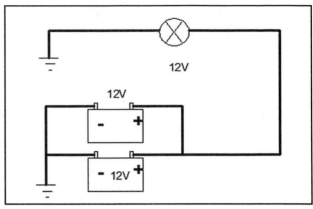

Two parallel-connected batteries allow the electrical system to deliver just 12 volts with high amperage reserves. This is a smart way to handle the high loads of audio/video systems.

Series-connected batteries work in an additive manner. Two series-wired 12-volt batteries deliver 24 volts, while three deliver 36 volts and four deliver 48 volts—far more than an automotive vehicle requires. Make sure you wire in parallel or you will have a meltdown in the vehicle.

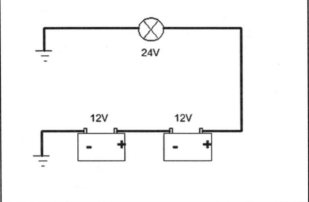

in series they add up to 12 volts of nominal voltage. If you join two batteries in series you will get 24 volts. More power, but you change the voltage and this can really damage circuits in a 12V system. Aircraft, industrial and many marine applications require 24 volts (or even 48 volts) and all electronics in those systems are designed to operate with higher voltage than a regular car or truck.

If you connect batteries in parallel you retain the 12 volts, but increase the available power. This is the way people running high-draw aftermarket audio and video systems, which can include 1,000+-watt amplifiers, supplement the vehicle's stock electrical system without overloading it. When these systems are not in use, the extra battery is getting charged just like the standard one. When extra amperage is called for, the second battery can supply it without burning up or short-circuiting the stock electrical system. These cars may also use a high-amperage alternator or two alternators.

The battery's purpose is to store electrical energy. Voltage is supplied by the alternator to the battery and the battery absorbs it until it is full and can

absorb no more. It is a chemical process and works in the reverse direction when the car is not running and needs electricity to start. When the terminals of a full or charged battery are connected to, say, a lamp, the lamp will light up. The chemical reaction is reversed and the stored electricity is released.

Electrical Voltage—A potential or pressure builds up at one end of a wire, due to an excess of negatively charged electrons. It is like water pressure building up in a hose. The pressure causes the electrons to move through the wire to the area of positive charge, always taking the path of least resistance. This potential energy is called voltage. Its unit of measurement is the volt. The battery is a warehouse for electricity, which is needed when the engine is running slowly or before it is started. In normal running the alternator supplies all the electricity required and usually with enough extra to recharge the battery at the same time. Once again, keep in mind that a battery is six cells connected together in series to give 12V.

The same is true with circuits, but in the inverse direction. If you connect two 12V lamps in series, the voltage drop will be divided between them, so that there is only 6V to each lamp. Therefore, they will not light up properly. The effect is the same as having a bad ground or corrosion in the circuit.

Connect them in parallel and there is 12V across each of the bulbs. Usually in vehicles we economize by wiring two lamps in parallel off a one-wire circuit. For example, with taillights, you run a wire to the first bulb and branch off in parallel to the second one, while the grounds for each bulb are independent. This saves the wire you would have to use by running two wires from the switch that

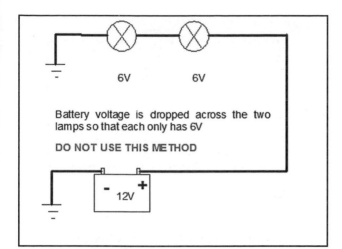

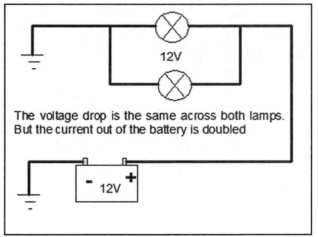

This diagram shows you what happens when the bulbs/lamps are wired in series. The voltage drops by 50% and the light glows weakly. You will get the same effect as if the connections were not made properly and have excessive resistance. (For example if the return wire mounting point to the unibody still has some paint residue under the ground wire, the light will not burn as brightly as it should, if at all.)

When you are wiring lights or other components that require a full 12 volts, make sure you wire them as these two bulbs in this simple circuit. Each component must get 12 volts directly, even though you can piggyback a single power wire to do it by splitting it before it reaches the components.

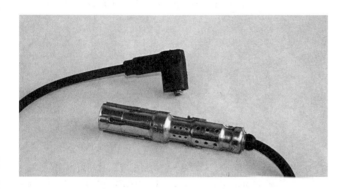

Have you ever tried holding the cap of a spark-plug wire while the engine cranks or is running? It has very high voltage of around +25,000 volts and will give you a big jolt. It is not dangerous however, because the circuit cannot deliver enough amps. As the old electrician's expression goes, "It is the volts that jolt, but the mills (milliamps) that kill." In other words, high voltage will make your muscles twitch, but high current (amps) will put your lights out.

controls the lights. This is the way to wire bulbs in parallel so they all shine brightly and can deliver their designated amount of light.

Problems to Avoid

There are two problems to be aware of with automotive electrics. The first is loss of electrical power, either partial or complete. The second is fire.

Most circuits in the car use 12 volts. That is 10 times less than electricity in the home and 10 times

less dangerous. You cannot get a shock from a 12-volt circuit. But vehicles work in a dirty and often corrosive environment. Connectors that are poorly made will become corroded and you will lose the circuit. Lights will dim and you can lose engine power. That is why good preventive maintenance requires you to verify the connectors on both the positive and negative sides of the battery at least on a quarterly basis. You will be surprised how much corrosion you will need to remove.

You can increase the standard voltage to as much as +40,000 volts without increasing amperage at all. If this were not possible it would be impossible for a spark plug to fire in today's high-technology engines, or even the muscle car V-8s of the '50s–'70s and it will give you a jolt if you hold the wire or spark plug as it fires. Because 12 volts is not a high voltage, it is not good at overcoming resistance. A spark plug gap is about all a 12-volt system can manage to jump and that takes a coil to do the work for it. By the same token, a pipe with a blockage in it will stop the shower draining. So if a car circuit has a poor connection because of corrosion or paint, the voltage to the component will be reduced and the lights will dim, the car will not start, the computer will not compute, etc, etc. (Bad grounds are often a problem after a car has been repaired from a major accident. Not all bodyshop workers may be aware that reattaching ground wires over paint can reduce or eliminate the ground.)

Even with today's improved batteries and better

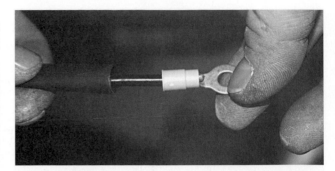

You can see where the paint was covering the metal where the taillight is to be mounted. There is also paint and visible primer on the inside of the actual bolt hole. This paint will eliminate any possibility of making a ground where the taillight mounts. That finger is pointing at the mounting stud for the taillight. This assembly is self-grounding through the housing, therefore it was necessary to clean out the bolt holes and scrape the metal clean where the stud/housing will seat on the roll pan to get a good ground.

Making extra grounds for the body, engine and instrument panel to the chassis (especially on a unibody vehicle like all modern cars are made today) is a great idea. Use 10ga. wires for a good ground. The photo on the bottom is the ground from the battery to the chassis when the battery is mounted in the rear of the truck bed or trunk. By grounding it close, directly to the frame below the bed, we only needed to run one long cable to the front for the positive side of the system. Make sure there is no paint under the ring connector to the chassis. This will prevent problems before they begin.

The most frequent cause of total vehicle loss of power is when the battery terminals get corroded and you try to start the car. The starter motor needs lots of juice, and it cannot get that power with corroded terminals. Therefore, the engine will not turn over. The round cell Optima battery is new technology and is far less sensitive to corrosion. They install like any regular battery, but they last longer and require less owner maintenance. Perfect for a race car, hot rod or any vehicle where the service is tough and maintenance isn't always given.

factory strategies to eliminate battery corrosion at the terminals, it is still a problem. Gasses from any acid battery will eventually seep out of the case and deposits will settle on the terminals. They will get a white powder on them and as that builds the battery will deliver reduced power until it quits completely. Clean your battery terminals at least every 90 days as part of a regular maintenance schedule. (Optima and side connect batteries are exceptionally well sealed and because of that corrosion will be less, but they still require maintenance.) Do it the same way when you are building circuits and check the grounds regularly. It is essential that every connector is installed

correctly or eventually resistance will build up and you will lose a light or something more critical. In computer-controlled cars, bad grounds are the cause of a large number of electrical gremlins.

There can never be too many grounds between the chassis, body, engine, transmission, computer and battery. You only need one power wire in the system, but extra grounds are a major form of preventing ground problems before they happen. When working on an older vehicle with rubber-insulated body mounts, you cannot expect ground wires to the body to produce good grounds, unless there are separate ground wires from the body to the frame. If the battery is grounded to the frame, make sure the connection has no paint under it and that the engine ground goes to the frame directly. Two engine grounds are always better than one.

I once had a friend tell me his engine wouldn't start and the engine was brand new. I looked at his car and told him to turn the engine over. It turned slowly and the full-size engine-to-chassis ground wire just about melted. The cause? The ground wire was connected to the engine properly, but he

Race sanctioning bodies such as NASCAR, NHRA and IndyCar have strict safety regulations in place for all classes of competition. Their rulebooks require electrical emergency safety shut-off switches. This disconnects the positive cable side of the system from the battery, stopping all flow. Since it is DC, it can only flow in one direction, so disconnecting the negative side isn't needed for safety.

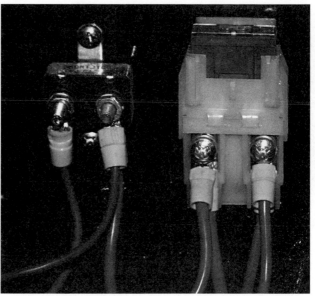

Here you can see a circuit breaker in the system right next to a 75-amp fuse, a main power buss, which is a multiple distribution connector that allows many wires to draw current from where the positive battery cable enters in the circuit.

had mounted that ground wire to the chassis at the motor mount. This particular motor mount was rubber insulated so there wasn't enough ground made to make it work. That is like having a wire too small for the job and it was the reason the wire got so hot. This was a simple fix, and the lesson here is to start with the obvious before tearing things apart to look for bigger problems.

Electrical Safety Cutoff—The positive side of the battery is the pressure side, which is why electrical safety disconnects on race cars are designed to cut the main power circuit right at the positive battery terminal. Once it is disconnected, all electrical circuits are isolated. If this safety shutoff would be connected on the negative side, it would still allow the electrical pressure into the circuits and if there was a short to ground it could result in a catastrophic vehicle fire.

Using Fuses—High-amperage fuses and circuit breakers are always wired into the positive side of the system by the OEM's and should be used when building any electrical system or modifying one. Place it near the positive battery cable if possible. It is always better to blow a fuse, or have a circuit breaker create an open circuit, than burn the wiring up and possibly the entire vehicle plus the house it is parked in. If you wonder how battery cranking amperage can be listed as 540 amps (above freezing) and the circuit protection is only 75 amps, it is because cranking amperage is available as "flash" amperage to get the starter motor turning the first instant of a rotation.

Spark Problems—Another problem created

when dealing with high current is that electricity can leap across space. When it does, the spark will damage whatever it contacts on both sides, from where it jumped to where it lands (each side where the spark bridged a gap). In the ignition system you will find at least one coil (to raise the voltage from 12V to +20,000 or higher) and from 1 to 8 or more spark plugs. The coil creates the voltage and the spark plug creates the gap required for the spark to jump. The metals a spark plug uses for both the center electrode and the wire it jumps to, the bottom electrode, may be simple carbon steel or high-tech metals like platinum or iridium. The more costly plugs with these metals allow the spark plug to operate efficiently in an environment where the amount of fuel the combustion chamber receives is the minimum amount required so fuel mileage is maximized. Very lean fuel mixtures require a stronger spark and high-technology metals in spark plugs are an aid to the spark.

You must also keep sparks away from any fuel vapors, so using the correct switches is an issue. If there could be fuel vapors (like when using carburetors or around a fuel tank) sparks are not a good thing. To combat the problem, relays are used because they require very little voltage to make the switch operate and the switch itself is sealed.

The switch on the dashboard that turns on the headlamps can be one of two types. It may close the circuit directly or, in some cases on today's vehicles, activate a relay, which does the actual switching. Using a relay, the headlight switch you moved only needs to carry a very small current, yet

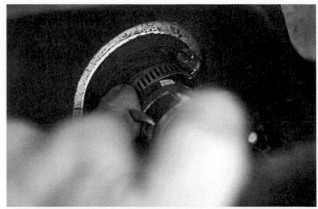

The spark plug circuit does not carry a high current load (few amps), but does deliver a strong spark to get the fuel in the cylinder burning. (High-performance ignitions with special coils can deliver more than 40,000 volts to the spark plugs, some systems deliver multiple sparks at low engine speeds to help modern lean combustion ratios continue to burn.)

Older vehicles use direct-acting headlights, so when you pull the headlight knob, it closes the circuit and does the actual switching right where the switch is located. This is fine as long as the circuit only draws 30 amps, but if the need is higher it is better to use a relay and lower the current at the switch inside the car. This can also lead to smaller components mounted in the dash, which gives OEM designers more space and places to cut weight.

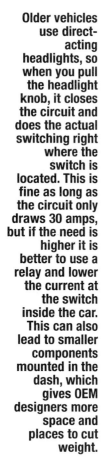

An additional grounding circuit that goes to the cooling system components will help fight electrolysis in the system. You need to ground the radiator as well as the heater core. The heater core is harder, but you can attach a ground wire using a hose clamp and then slip the wire under it to hold it in place. These two photos show how it is done. The first is where the hose clamp goes and the second is just to show how it should look after the ground wire is in place and the clamp is tightened. (This is not the heater core pipe, just an example.)

it will turn on the headlamps in complete safety, even though they draw a lot of current.

Electrolysis Problems

Electrolysis occurs where there are two different metals joined together with a circuit running between them, such as in a cooling system or a ground wire to the chassis or body. What happens is that the metals start to erode/corrode and eventually you will lose the connection, or inside the engine the heads and intake manifolds will start getting eaten away because of electrolysis. This is why copper ground connectors mounted to a steel frame or chassis connection can corrode more quickly than if the battery cables were steel. They need periodic inspection because of this problem. Some older cars had a positive ground (yes that is possible with DC), however, electrolysis is greater with that polarity so modern cars have negative to ground. Interestingly, you will find aluminium components that are in the cooling system (thermostat housings, radiators, etc.) can erode from the same cause. It seems that a cooling system with dissimilar metals generates its own electricity (up to about 1.5 volts —almost like a battery) and needs a good ground to help combat this process.

Keep an eye out for hot circuits like this air bag impact sensor while inspecting the car. You do not want to touch them unless the positive and negative battery cables are disconnected. Air bag system components have blue and yellow wires to help you identify them. You wouldn't want to trigger the system accidentally, inflating the air bags and doing a lot of damage inside your vehicle. Note also that this sensor circuit has a locking system using plastic loops and ties.

In this chapter we will learn where the major components of the electrical system are located on a vehicle. This will include the battery, fuse box, starter, alternator, heater—A/C, engine fan, windshield wiper motor, electric fuel pump, ignition system, the turn signal/emergency flashers and a few other items. Although this will probably be very basic for most readers, we cannot simply assume everyone knows how to locate and identify these components.

For example, do you know where the starter is located on Cadillac Northstar engines? If you didn't know and were looking for it, you could waste a lot of time and never find it, because it is actually located inside the engine under the intake manifold. Although this is the most unlikely starter location on domestic vehicles, there may be others on imported vehicles that are also different.

MAJOR ELECTRICAL COMPONENTS

Sometimes the function of the component will dictate roughly where it has to be located. For instance, since the starter motor has to engage the flywheel, it will be down low and at the rear of the engine on rear-wheel-drive cars on the left side (driver's) on typical 4-cylinder front-wheel-drive engines. Since the alternator is powered by the fan belt, it will be near that location in the front.. There are variations on these themes, but always start looking in the usual places especially with American cars as they are almost always in the same places.

Note: Read this before you touch anything with light blue wires and yellow connectors. It is a crash sensor for the supplemental restraint system, usually identified inside the car with the initials SRS on the steering wheel and/or dashboard, and commonly known as the air bag. Almost all vehicles since the early '90s have had air bags to protect drivers and passengers in the front of the car. The components that make them inflate are color-coded with light blue wires and yellow connectors. They may be in front of the car under the radiator support or at least in front of it. They can also be in the back under the rear bumper and on some cars they are located in the doors or along of the sides of a car (at the bottom of the sill), to activate side-impact air bags. Regardless of where you see yellow connectors with light blue wires, before you touch them remove both positive and negative battery cables from the battery. This is important for your safety and the safety of your car. Triggering an air bag by accident gets very costly as this usually destroys the dash as well as the air bags.

Battery Location

The starting point for anything electrical in any vehicle is the battery. It is usually located on one side of the engine compartment. Custom cars, race cars, hot rods and compact sport sedans generally move it from the front to the rear of the vehicle. When they are in the trunk, or bed of a pickup truck, they are always mounted in a protective box or

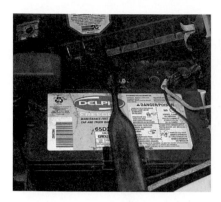

The battery is usually located on one side in the engine compartment, unless it is a custom or someone is looking for better cornering by moving front-end weight to the rear, such as in the bed of a pickup or the trunk of a car. When it is in the trunk it is mounted in a protective box before mounting.

The engine compartment in today's vehicles is pretty cramped. There are hoses and a lot of other clutter to deal with.

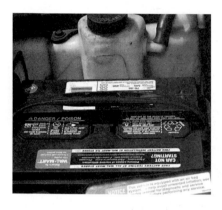

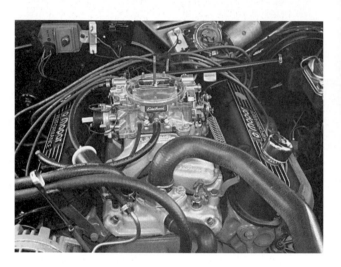

If you find a battery with no cables mounted on top, then it is likely a side-mount battery. These are commonly used on GM vehicles.

Muscle cars from the '60s to the '80s have a lot of room in their engine bays. You can get at everything from the top and they are far easier to work on than today's cars and trucks.

mounting so they will not move around.

You will generally find positive and negative battery cables mounted on top. But, don't be surprised if you only see one battery cable going forward inside the vehicle or under it. It is better to ground the battery to the chassis or unibody at the trunk or bed to save weight and to minimize the risk of having a cable damaged.

Some batteries mount the cables on the side to reduce post and cable corrosion. The cable mounting is with a 5/16 headed mini bolt so get the right tool or you will round it off and then face a lot of work to remove it and get another bolt to fix it. Most GM vehicles use this system, but when replacing the battery some people convert to top-mounted posts and new cables.

A typical engine compartment in a modern automobile can be a cramped maze of wires, hoses, tanks, belts and as many engine components that can be stuffed under the hood. Working on newer vehicles requires patience and keeping track of where everything goes, before you remove it. Mark

each hose on the end you remove as well as the place it was attached with the same name or number. By doing this you will get things back together correctly.

Few vehicles manufactured after the mid-'70s use point-type distributors for the ignition. Now they are all computer controlled and generally last 100,000 miles before needing maintenance (plugs and every now and then a coil wire). To find out what is wrong on these OBD II vehicles it is necessary to plug in a scanner to the OBD port and check fault codes.

Many muscle cars have large, spacious engine compartments, and they are easier to work on. They are big and everything is generally accessible from the top. In most cases you can lean over a fender and change plugs, plug wires and tune the

The alternator on top is a chromed OEM-type mounted on a Chevy 454 custom engine. The alternator is accessible and representative of almost all alternators regardless of make on V-8s. Chrysler alternators look a bit different but are also mounted on top of the V-8 engine. The MSD is an example of a performance aftermarket option that delivers higher output and a custom look.

carburetor. Everything is right at hand, except for some distributors, which can be in the back behind the carburetor. If they have electronic ignitions, the only maintenance items on a distributor are the plug wires, cap and rotor. Ford vehicles used a front-mounted distributor on all V-8s starting with the FE big blocks as well as the small blocks. Chapter 7 deals with how to eliminate points and still keep the engine compartment looking 100% stock.

When you start from scratch you can dictate how much room will be in the engine compartment. On the 1956 Ford pickup we used in a how-to chapter, the chassis was modified and the body placed in position before the front clip from a Camaro was welded in place. This determined the engine location and allowed the builder to leave room between the distributor and firewall. When you are

building your own project car, consider carefully where you will locate your electrical system components.

Alternators

Aftermarket alternators, from stock to custom, are available for any vehicle. High-output, high-amperage alternators are always available for heavy-duty applications. Although you may need to modify the mounting on some units, most high-amperage alternators are actually built from stock housings so they fit just like the OEM unit you are replacing. If you are going to a custom unit like the one from MSD, you will find that they have a mounting system that will fit your particular engine and vehicle requirements.

Calculating Amperage—If your car needs 75 amps to start, get an 80-amp alternator to make sure you have some margin of error. Likewise, you need to know the total amperage needed during day-to-day operation of the vehicle. Some vehicles need more than 100 amps if they have a lot of added aftermarket items. They must have at least an extra HD battery plus a 110-amp alternator. You cannot go wrong replacing your OEM or present alternator with another of the same specifications if it has been doing the job. Be sure to make the battery(s) and alternator with the same amperage rating. But, if you have been finding yourself short on voltage (the lights flicker when an accessory is turned on at night is the first clue to most drivers), recalculate your actual total amperage requirement by adding up the requirements of each accessory. The total should not exceed the amperage of either the battery or alternator. This includes lights, audio and video systems, air conditioning, navigation GPS, power steering (very few are electric at this time), heater and fans. If you have an electric cooling fan you need to add it in also. If they do, you need to upgrade both your battery(s) and alternator. High-amperage alternators generally do not last as long as OEM because they produce far more heat. Remember, two batteries are better than one high-amp alternator and cost less.

Identifying the Alternator—When you open the hood on a vehicle you will see at least two things driven by a V-belt or serpentine belt system. Those things could be an alternator, power steering pump, air conditioning compressor or some kind of accessory like an air compressor, and the engine fan. All vehicles will have an alternator and power steering today, but A/C is still an option. You must know what these things look like and how to identify them.

Generally an alternator will have some sort of

This is a Chevy pickup A/C compressor, not the alternator. Notice that it is solid (no slots for air to go through) and it has a strange-looking electric clutch drive. A/C compressors are also distinguished by their refrigerant hoses, something alternators will not have.

The alternator in this picture is for the same Chevy truck at left, but it is mounted low on the side of the engine with access from the bottom only. This is becoming more common on modern cars and trucks.

Since the 1950s some electrical controls have been in the steering column. Turn signals, horn, and others are all running through the big connectors on the column. An important system is the neutral/park safety switch that keeps the vehicle from starting unless the gear selector is in neutral or park. On vehicles with a clutch there is a controller that will not allow the car to start unless the clutch is pushed all the way down to disengage the engine from the drivetrain. (The clutch control switch will be mounted above the clutch pedal and will probably look just like the brake switch.)

This is where most vehicles have their fuse panels mounted if they are from the '50s through the '80s. In most cases it is the best place to locate the fuse panel to keep it out of the way and out of site, without making it tough to work on.

cooling fan on or in it and a case that has a lot of open vents for cooling. A/C compressors have an electric clutch at the front that activates the compressor. They never have cooling spaces. But, alternators do not have any hoses or hard lines going to them either, just wires, unlike the A/C compressor or power steering pump.

Power steering pumps, like A/C compressors, have hard line pipes and/or hoses attached, but the power steering pump doesn't have a clutch in front or any wires. The hydraulic reservoir usually will be mounted on the pump, but it can also be in a remote location. The alternator will normally be mounted near the top of the engine, but not always.

The starting point when wiring any vehicle from scratch is generally a wiring system purchased from a supplier and then adapted to the selected vehicle. The builders of the custom '56 Ford pickup illustrated throughout this book selected a kit and

the fuse panel it comes with. All they had to do was read the instructions to install it. If you want to do it all yourself, you can either replicate the stock wiring harnesses, if you have them, or create your own.

I don't believe it is feasible or practical to start from scratch and make your own harness. It will cost you more money and take more time than using a wiring kit. I've installed wiring kits many times and also have created harnesses from scratch; the kits are a much better way to go. Even if you only use the basic vehicle wires without accessories, you simply cut the ones not used and terminate them with a cover of shrink tubing.

Locating the Fuse Panel

The first step in installing a custom wiring system (homemade or purchased) is to decide where the fuse panel will be mounted and where the exterior wiring will leave the passenger compartment. The fuse panel is generally located under the left (driver's side) side of the dashboard, but it can also be outside. If the car/truck was originally made for the Japanese market you may find everything on the right (passenger) side.

A lot of electrical circuits in cars from the 1960s to today's vehicles are run through the steering column. Headlights, parking and directional signal lights, flashers, radio, transmission selector, high beam switch, etc. may all connect from the fuse

The connector provides an excellent location for trouble-shooting problems with any of the systems in the column. Just pop it apart and you can check both sides of the connection. Use a multimeter (see page 29) to test for continuity, power or resistance.

On race cars the fuse panel should be mounted on top of the dash, near the passenger window, so fuses can be checked by a crew member without climbing through the roll bar system or go under the dash for a simple electrical fault.

GM likes to hide fuse panels in their late-model cars and trucks. This is the fuse panel in the Chevy pickup shown in the photos throughout this chapter. It is mounted on the left side of the dash and is covered when the door closes. The close-up shot shows the relationship between the panel and the door switch that shuts off the dome light.

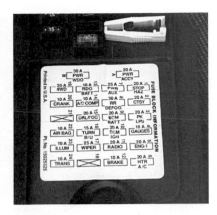

Inside most fuse panel covers you will find a schematic showing where each system is fused and how big the fuse is. Make sure to follow this guide during repairs.

Our Chevy truck has the power distribution buss separated from the fuse panel. It is outside the cabin in the engine compartment, mounted to the inner fender of the vehicle.

panel to a steering column connector. A big connector usually allows removal of the entire column or the wiring harness from the vehicle without having to dive into the column itself. This is also a good place to test circuits so you know if the problem is in front of, or behind the wiring going to and from the column harness.

Race cars should have the fuse panel visible and accessible, so designers usually mount them on top of the dash or in an accessible place in the engine compartment and run the wires down to interior and exterior connections.

Sometimes the fuse panel (and power distribution system) are hidden in strange places, between the door and dash (on the driver's side), under the dash, in the engine compartment, on the passenger's side or other locations out of sight. Try to get a service manual for your project vehicle as it will make your job easier.

Always look inside the cover of a fuse panel

(inside or outside in the engine compartment) for a map that will tell you which fuse goes to what system as well as the amp rating of the fuse.

Power Distribution Circuits

Every vehicle has a power distribution circuit and while some are combined with fuse panels, some are stand-alone systems and can be mounted anywhere, even on the power takeoff side of the starter motor.

Under the hood of our 1997 Chevy pickup, you will find the power distribution center on the left side fastened to the inner fender. It has a plastic cover for protection from stray wires or other possible foreign-object damage. Remove this cover and you will see the power distribution buss circuit of the pickup. All it does is use a big wire from the exit side of the master power fuse and allow other wires to get the voltage they need to operate. The large wires are from the battery/alternator and the

Here is a 100-amp fuse for the starter and lights for an Acura Showroom Stock racer, plus the others from 50 amps to 20 amps.

This is the power distribution system of a custom Ford pickup. Note how clean and simple the installation is.

smaller ones go to lower amperage circuits. All modern vehicles have a fused starter circuit. The starter on older vehicles may not be a fused circuit, but if you are building a custom system, make sure it is fused.

Look at the power distribution system on the Acura race car shown in the photo above. The battery terminal in the photo is actually the ground cable, while the positive wires go from the removed positive cable to the panel on the left of the ground terminal. (To bulletproof the electrical system all grounds end up at the battery, eliminating all chassis grounds that can fail due to corrosion or driver-induced crunches.)

The big fuses in the Acura's panel are rated at 50 amps and the small, white, flat fuse on the left side near the big wires is rated 100 amps for the starter and lighting system. The rest of fuses are for lower-amperage circuits. All of the light switches and other driver-controlled items are inside the car on a switch panel between the gauges and the driver.

Starter Solenoid

In the power distribution system used on the '56 Ford pickup shown in the photo at above right, the positive battery cable drops off the power to the big starter terminal, but it cannot operate without the fuse on the right in this photo. The starter is being used as a junction block at this point, then the power comes up to the right side of the 75-amp fuse. From there everything goes through the fuse before it is sent to power the vehicle and starter solenoid controlled by the ignition key switch. All of the interior power goes through the 50-amp circuit breaker to the left of the big fuse.

Remember that a solenoid is a type of mechanical relay, so it is the small wires coming back to the starter solenoid that makes the solenoid/relay kick the starter gear into the flywheel and turn the engine over. Power goes directly from the circuit breaker to the fuse panel inside the body before

This is a solenoid that can be used for many things. In this case it is being used as an electrical shut off, but take off the wires and you have a Ford-style starter solenoid.

being sent back out to each individual circuit. Not all systems are as simple and easy to diagnose when problems arise.

There are two big terminals on a Ford starter solenoid. If you had one in your hand the one on the left would be the positive battery cable connection and on the other is a big cable that leads directly to the starter motor. No electricity can go through the big cables to the starter unless the solenoid is activated by a low voltage signal from the key ignition switch. Once power from the key ignition switch hits the small terminal also on the left, the solenoid closes and power goes directly to the starter. Accessory circuits in the vehicle can take power from the small right terminal.

A Chrysler starter solenoid looks different, but works similar to the Ford-style solenoid. Power goes from the positive battery cable to the starter and then back up to the solenoid on the fender. Power is delivered to the inside fuse panel through the fusible link. (That is just an in-wire fuse to protect

These two photos were taken as a new solenoid was being installed on a 1972 Barracuda. In the first photo you can see how it looked before we replaced it with a new unit and some new harnesses. The old one had some nasty looking cables going into and out of the unit. It also didn't have a fusible link in the circuit or a fuse of the correct amperage. We wired in the new one and connected the fusible link into the circuit going from the solenoid to the interior of the car.

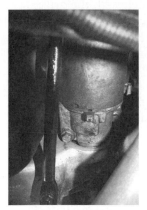

The first shot shows a starter installed in a vehicle taken from the bottom looking up. You can see that space is at a premium in the bottom of the pickup truck as well as the top. Even though it is hard to find, remember it will always be connected to the rear of the engine or the front of the transmission. The second shot is of a high-performance replacement starter with high-torque gearing to make a small starter motor strong enough to get the engine turning fast. This type of starter is used on Chrysler and GM vehicles. (This is for a Chevrolet.)

all vehicle wiring with the exception of the starter circuit.) When the starter switch is turned to start by the key (two right angle connectors on the solenoid), 12 volts go to the cranking solenoid on top of the starter motor and make it kick the starter gear into the flywheel of the engine.

Distributor

The distributor on most GM and Chrysler V-8s built from the '50s through the '80s is located at the rear of the engine, distinguished with 8 plug wires attached to the cap. The late-model GM HEI system (1975 and up) has an integral coil as part of the cap assembly, so do not look for a coil wire. If it has a regular cap and coil wire going from the the

center of the distributor cap to a coil, it will be a standard points ignition or it has been modified to look stock yet be an electronic ignition. The first HEI distributor was introduced in 1974.

On all V-8 Chevy engines, the distributor is at the rear of the engine, toward the middle and very difficult to reach. The engine used in the custom '56 Ford pickup we've discussed is a big-block Chevy (454) with an MSD HEI distributor and 8mm spark plug wires. The owner protected the plug wires by running them in protective mounts to avoid heat damage and to keep them out of the way and organized. Even on the electronic distributors such as an MSD unit, the cap and rotor still need to be changed after about 50,000 miles for maximum performance.

Heater Blower

The most basic heater blower motors generally have three wires: hot lead low speed, hot lead high speed and ground. On vehicles from the '50s and early '60s the heater blower motor was on the firewall somewhere, while inside the vehicle was where the heater core and fan assembly (squirrel cage) was located. There are also doors that control airflow to the heater for the cabin as well as the defroster ducts. Later model vehicles have far more complicated fans and heater and A/C boxes. To find the heater core and/or blower motor look for the water hoses from the engine going into and out of the firewall. Behind that point you will find the heater core. To find the fan motor, turn it on high and look for where the sound is coming from, or if that doesn't work, look in a shop manual.

Power Window Mechanisms

Power window wires, mechanisms and motors are

The distributor on most older GM (left) and Chrysler (middle) engines is mounted in the top rear of the engine, making it difficult to work on. Older Fords (right) have the distributor up front.

Heater blower motors can be located through the firewall as on this '56 Ford truck, or inside the vehicle somewhere under the right side of the dash. If you locate the heater core inlet and outlet pipes, you will know to look nearby (under the dash) for the blower motor. Some GM and Ford vehicles had the blower motor and heater core outside in the engine compartment for space reasons.

Looking at the beautiful custom dash of the Ford pickup, it might look like it is complicated, but it is really simple. The billet aluminum insert was purchased on the web. Pro Street guys love to have a big tachometer on top of the dash, but most people would find it distracting and that it limits visibility. Having both instruments in the insert with just two other gauges is a better placement.

located inside their respective doors. The power switch will be in the door panel, with the driver having switches to control all power windows. In the photos at top right on the next page, the windows are being installed in the doors and the wiring was run into the doors and to the motor. There are just three wires on most vehicles— ground, up and down. If there are also door locks they would also have a switch on each door even though they will lock or unlock all doors. Repairing these systems is easy, but the hard part is getting the door panels off of factory doors. The window switches were located in the center console crafter by the owner and the door switches were mounted under the dash on either side of the vehicle. Once outside the vehicle, a remote operates the doors.

Shifter Safety Switches

Earlier automatic transmission-equipped cars have a park/neutral safety switch located somewhere within the shifter mechanism. This switch prevents the engine from turning over unless the transmission is in park or neutral. On later-model cars it is part of the computer system, so before you start working on something make sure you know which wires are to what systems. Make sure you always wire this switch into your vehicle if you are making modifications or building something custom with an automatic transmission. On manual transmission vehicles, wire in a clutch switch, like the brake switch, into the starter circuit so the engine will not start unless the clutch is fully depressed, even if it is in neutral.

Switches

On the Ford and Barracuda, as on all headlight switches, the units control all lights on the vehicle allowing you to just turn on the parking and side marker lights in the first position and the headlights in the second. On most new vehicles, the high and

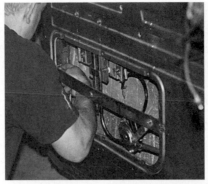

The Ford dash insert on the previous page has custom Stewart Warner gauges. In the photo above you can see how they all fasten in the back. The lights are already installed in each instrument and speedometer.

Electric windows from the factory or aftermarket generally operate the same, albeit with less control on the OEM systems as the computer is involved in the OEM circuitry. You will find that the window motor will have three wires: ground, window up and window down.

An aftermarket remote system for doors and windows is shown here. From top left to bottom you will find the remote units, control unit, relays (center) and master wiring harness. OEM units contain the same things but you need to locate them using a wiring diagram and/or shop manual. Many OEM systems use the computer of the vehicle also.

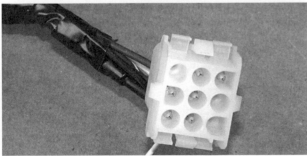

The connections are made with simple plugs. By using a male and female connector (female shown), the insert slips into and out of the dash with ease leaving only the direct reading gauges for oil pressure and water temperature to be removed if you are doing more than just changing bulbs or color. Chapter 11 includes step-by-step installation of this cluster, beginning on page 155.

low headlights are controlled with steering column-mounted controls. All of these headlight switches have a large connector that plugs all of the wires required into the switch. Some early cars (1940s and 1950s) had each wire screwed into the head/park light switch.

Some vehicles no longer use bulbs for the high beam and turn signal indicators. The directional and high beam indicator lights on these dash inserts are actually LED lights and they generally last the lifetime of the vehicle. If you are building a custom dash insert, spend a couple of extra dollars out front and you can forget about them not working later. If they quit working you know the outside lights or flashers are bad.

Wiper Motors

Windshield wiper motors come in many sizes and shapes, but they all look similar to the unit in the photo above right. While this one is easy to access, many are under cover panels or air intake grids in the cowl for the blower motor, so refer to your service manual. The good thing about wiper motors is that they generally last the life of the vehicle. If you have a bad one, don't be afraid to get one from a salvage yard or a rebuilt.

This custom shifter was added to the vehicle to get the shifter off the steering column. It has a safety circuit that will keep the engine from turning over unless it is in park or neutral on an automatic transmission vehicle. The manual transmission cars can require that the shifter is in neutral and the clutch is pressed all the way. Never fail to install these circuits as they are critical safety systems. Just connecting those two wires can make a big difference. Other circuits can also be connected to this switch.

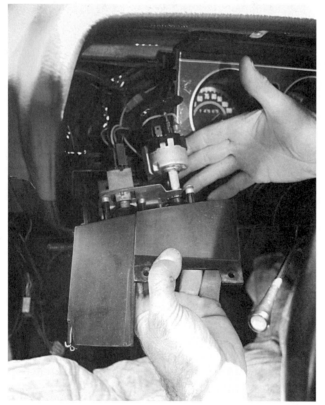

The heater fan switch and the dome and instrument dimmer part of the headlight switch on the 1971 Barracuda are being removed. The same panel also holds the heater controls. The dimmer switch allows the owner to adjust the instrument lights while moving and to turn on or off the dome light before opening the door.

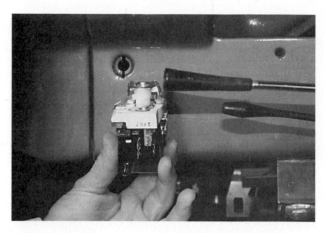

In this custom '56 Ford pickup, a 1980s GM headlight switch with an integral dimmer function to adjust the dash lights and the dome light was installed. (You can turn on the dome light without opening the doors on both the pickup and Barracuda.) The front part of the switch assembly is where the dimmer is located, using a ceramic-mounted coiled rheostat to do its work. Most modern vehicles have a wheel-type control to adjust the lights and turn on the dome light, but they are just a variation on the coiled rheostat circuit.

At left is the backside of the original foot-operated headlight dimmer switch from a '56 Ford pickup, found on all cars of that era up through the '70s. At right is the part that sticks through the mounting plate that screws into the floorboard. These switches were the standard in the '60s and '70s, but dirt, moisture and even snow would eventually corrode them, which is why all dimmer switches are now mounted on the steering column.

Emergency Flashers

On some Chevy pickups the flashers are located in the middle of the dash, under several panels that must be removed to change them. This is at least a 45-minute job for the experienced. On custom vehicles they are almost always on the fuse panel. This is just one of many reasons to have a service manual for your vehicle. The factory no longer creates them for major markets as electronic service manuals are less costly and easier to distribute from the company's database.

Wiper motors are usually mounted on the firewall or in a recess under the cowl under a cover.

Dash lights for custom instruments such as installed in the insert of the '56 look like this. They pop out easily to change them and you can even change their color by swapping the little plastic bulb/cover (red or green). The three wires in the right photo attached to the insert next are LED's for the turn signal and high beam indicators.

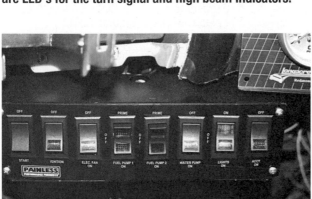

Flashers are used for two circuits in vehicles—regular turn signals and emergency flashers. Both of these circuits are generally located inside the vehicle.

This switch panel is installed in the Acura race car below the instruments between them and the steering column area. All switches are in easy reach of the driver. They give the driver a lot more control than the stock wiring system, even though some of the original wiring was maintained because of rules.

Electric Fuel Pumps

Electric fuel pumps are used on a lot of custom and race vehicles as well as on OEM stock vehicles. The difference is in the size of the pump as well as the pressure and volume it is capable of delivering. They are all mounted in the back of the vehicle, but OEM units are generally inside the gas tank, while aftermarket ones are in-between the gas tank and the outlet line going forward. Stock systems mount the electric fuel pump inside the tank connected to the standard tank float and rheostat that tells you how much fuel is in the tank. They are hard to work on and many times require dropping the tank

to gain access to the pump.

Knowing where components are located and what they look like will give you a step-up in solving and repairing problems in electrical systems. Don't be afraid to take things apart to see how they work. But if they are complicated, make notes on how they were installed and how to get them back together. Patience is one of the most important assets for automotive work and general mechanical building and repair. The more experience you get the easier big projects will be.

This F.A.S.T. pump system is externally mounted from the tank and comes with the pump, filter and circuit breaker as well as a fuel pressure regulator that works with both carburetors and electronic fuel injection (EFI).

It doesn't matter if you are working on an old car, a modern one or something like this Mopar show car on display at the New York Auto Show, you need to know how to identify components and where to get the electrical information you need.

Chapter 3

Understanding Electrical Loads

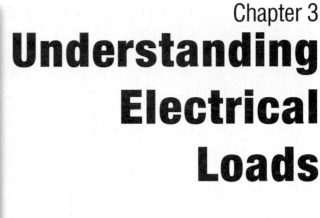

When you find yourself with wires all over the place and are about to give up, just remember there are tools and suppliers that can help you through anything. If you need help calculating loads on circuits before you create them, the suppliers that made the components you want to install can help you. Here you see one of the techs at F.A.S.T. / COMP Cams showing how to install one of their EFI systems. He is also there to answer your questions about their EFI systems and the electrical loads it might impose upon your vehicle.

Why is it important to know what type of load is being carried by the circuits and power supply of your vehicle? Because these loads come from the components installed by the factory and aftermarket components you may install. If they do not match the capacity of the stock electrical system you must make changes to make sure your vehicle starts up and runs correctly in all weather and driving conditions. This chapter will give you an overview of the basic automotive electrical systems (more in-depth information is in Chapters 6 and 7), the loads they carry, the different types, why it is important to know which is being carried on the circuit, and how to protect against abnormal peak loads. It will also explain why you must calculate the amperage draw the entire vehicle may have at maximum load requirements.

Knowing the total load is especially important when it comes to adding aftermarket electronics, such as megawatt stereos, DVDs, nav systems, etc. It wouldn't take much to overload the vehicle's electrical system, which is already at or near its limit servicing all of the basic electrical/electronic systems. There is always a solution even to the point of adding an extra battery and a second alternator, but loads are the most important piece of information you need. So, do not find yourself by the side of the road because you turned on the headlights while the radio was beating out 2,000

watts of bass from the stereo. Instead, verify before you fly and make sure your electrical system is adequate with all the accessories installed by the factory plus the ones you installed after the vehicle left the factory.

Resistance Loads

You should be familiar with the concept of electrical resistance after reading Chapter 1. We know that wires must be rated to match the load. If you install a wire that is not rated high enough for the load, the wire will overheat. An incandescent lamp, the type traditionally installed in homes and vehicles, is just a piece of wire that gets hot because it has resistance. When it gets hot enough it glows or incandesces. A 60-watt bulb will produce more light than a 15-watt bulb and has a higher resistance. They also both produce a lot of heat. So the resistance wire inside the bulb is specially made for the application—it gets hot without burning, it is tolerant of vibration (whether hot or cold), and gives yellow light when heated.

Resistance loads are relatively simple and do not vary. Other types of loads can change their impact on the overall electrical system depending on what is happening. Other examples of resistance loads are adapter locations for computers, cell phone chargers and iPod transmitters (that go in the 12v receptacle that used to house a cigarette

This power distribution buss in a Chevy pickup delivers a lot of voltage to each circuit. At one time there might have been only 4 circuits needing power, today the number is well over 20. The increasing loads on circuits create new problems in generating and storing electrical energy to feed them. There are many new automotive systems that impose large power drains for short periods of time, such as electric power steering. These also need to be addressed if you are installing them. But, that doesn't necessarily mean just increasing the power supply. If you are using an ECU for engine management and other duties, you must protect the unit from sudden changes in voltage, which can occur when other high electrical load systems are switched on.

This is the typical headlight like those used on older vehicles (rectangular or round). Older OEM versions used incandescent bulbs, but the move in both the OEM and the aftermarket is to use more efficient halogen bulbs. These new types of bulbs also demand more electricity, creating more resistance. The amount of wire that can be stuffed inside a bulb is obviously the major limiting factor, which is why some filaments are double and triple coiled. The larger the filaments, the more amperage they draw, something to remember when adding aftermarket bulbs.

Modern headlights on vehicles are very high tech. They can be formed in any shape and their reflectors aim the projected light away from the eyes of oncoming vehicles. But, they also demand more amperage/electricity through the wires feeding them. That is why retrofitting these types of lamps can require a rewire to a larger wire gauge over the entire circuit that can handle extra amperage loading.

Diode bulbs are ideal for taillights because they are an almost-instant-on light and can be built into clusters so if one goes out the rest continue working. These bulbs use less electrical energy and draw less amperage, added benefits.

and taillights to turn signals and interior dome and dash light.

Diode Lamps—Diodes produce an intense light from a small point and turn on instantly, so they are great for brake lights when used in a cluster. Also, because there are lots of them grouped together, if one fails you still have the others.

Magnetic Loads

The ignition system, starter motor and cooling fans are all examples of magnetic loads, which is quite different than the resistance loads generated by lamps. The load from a starter motor comes from generating a magnetic field that is used to do work by applying it to a magnet (solenoid). The motor then turns and the starter drive gear engages the flywheel and turns the engine. Its load though is real and will draw current from the battery, lots of current (as much as 40 to 75 amps in flash load). Although you can test a bulb to find out its resistance and from there determine its load, you cannot test a starter motor and make the same calculation. But, the manufacturer of the starter motor will have rated it with a specific amperage draw and this information is available.

Starter System—The diagram on the page 25 shows a typical simple starter installation circuit. The ignition key activates the relay unit, which in turn sends power to the solenoid. The solenoid is

lighter). These receptacles draw significant current, but they do not have any special requirements above the power they draw. The most common type of resistance loading in a vehicle comes from the lighting systems used on vehicles, from headlights

used to close the starter motor circuit and to push the starter gear forward to engage the flywheel teeth. The fuse for this circuit should be at least 75 amps and located somewhere near the starter. It may be located in the exterior fuse box along with other fuses, on the firewall or on the inner fender. The key things to note about a starter system are:

• Starter motors use lots of current and need to have a large supply with positive power and ground wires that can carry as much as 75 to 100 amps.
• They must have a fuse in the line.
• The large current requires the use of a separate switch close to the motor to supply the current. This is the solenoid arrangement shown at right.

Remember, the higher the load, the more amperage required, which means larger wires and connectors in the circuit. I have seen a combination bad starter and engine ground, cold weather and thick engine oil make a starter circuit draw over 140 amps. Needless to say it blew out the fuse and fried many wires.

Windshield Wipers

Windshield wipers utilize two mechanical devices to perform their task: an electric motor and worm gear provides power to the wipers, and a cog drives a linkage to convert the rotational output from the worm gear into the back-and-forth motion of the wipers. It takes a lot of force to accelerate the wiper blades back and forth across the windshield quickly. In order to generate this type of force, a worm gear smoothes the output of a small electric motor. The worm gear's reduction can multiply the torque of the motor by about fifty times, while slowing the output speed of the electric motor by the same amount. The output of the gear reduction operates a linkage that moves the wipers back and forth. Inside the motor/gear assembly is an electronic circuit that senses when the wipers are in their down position. If you turn off the wipers the circuit maintains power to the wipers until they are parked at the bottom of the windshield. It then cuts the power to the motor.

This circuit also controls the wipers when they are on an intermittent setting if they are a modern system. A short cam is attached to the output shaft of the gear reduction. This cam spins around as the wiper motor turns. The cam is connected to a long rod; as the cam spins, it moves the rod back and forth. The long rod is connected to a short rod that actuates the wiper blade on the driver's side. Another long rod transmits the force from the driver-side to the passenger-side wiper blade.

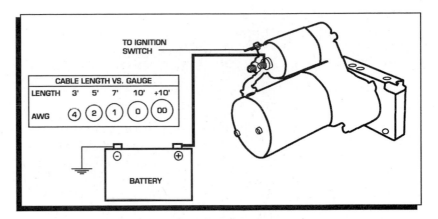

This starter system has large cables going from the battery to the starter and ground. The longer the distance to reach the starter, the larger the wire will have to be. For example, if the distance is three feet, then a wire gauge of 4 would be recommended. Gauges go all the way down to 00.

This small fuel pump will deliver enough fuel for nearly 400 horsepower. It is quiet and can be installed easily as it just goes from the fuel-out line of the fuel tank, to the fuel-in line of the injector fuel rail. There must be a fuel filter before and after the pump to make sure no junk gets to the injectors as well as a return line for bypassed fuel to return to the tank.

What, you ask does this have to do with loads? The answer is simple. The system might be rated for 20 amps with fuse and wire sizes to carry that 20 amps load, but what happens to the amperage draw when the wipers are stuck to the window by ice, when something gets into the linkage and keeps it from moving, etc? The system may actually try to draw 30 to 40 amps and burn out a fuse or the wiring if the fuse doesn't do its job. Once again it is the load as designed vs. the actual loads that can cause serious problems.

Engineers always try to provide a bit of extra safety margin in their circuits when they are designing them. But in today's automobile manufacturing, every bit of over-engineering adds cost to a vehicle, which is discouraged.

These two pressure switches are like the ones used by GM so that when there is no oil pressure (the engine is stopped) the electric fuel pump will not have power. These switches can be purchased in pressures from 10 psi to well over 50 psi.

This Holley Commander 950 ECU will handle up to 500 hp with the correct injectors and manifolds installed on a high-horsepower engine. As part of a complete kit, it delivers excellent performance on street engines with Throttle Body Injection (TBI) and can replace a carburetor with EFI in less than a full day's work. Holley also has a system for more powerful engines. How much amperage does this system require? It uses a 40 amp relay. That's a lot of amperage for electronics.

Electric Fuel Pumps

Although electric fuel pumps themselves are easy to understand, the circuits that control them are a little more complex. All pumps have a fuse in the circuit, but more control is needed than just overload protection. This is especially important when controlling a highly flammable liquid—like gasoline. Most pumps are powered during cranking and will continue to run after the engine starts. If the ignition is turned on without starting the engine, the pump only runs for a second or two. This safety measure prevents fuel from being pumped into the engine if an injector is leaking. It also prevents fuel from being spewed all over in case the vehicle is in an accident. How much load demand in amps might these pumps generate? That all depends upon the type of fuel delivery system you have (EFI or carbureted), how the system bypasses extra fuel that isn't needed and the maximum pressure required at peak demand.

All of these factors can cause demand and amperage draw to change dramatically in a matter of seconds. In the time it takes to go from idle to wide-open throttle (WOT), the load could easily go from 10 amps to 30 amps. If the electrical system isn't designed for an EFI system and you add one, you will need to resize the wiring and fuses in the circuit to the new demands.

A computer-controlled relay is the most common method of controlling the pump. The computer looks for an rpm signal to decide if the engine is running. If it sees the proper signal, it grounds the relay, and, in turn, powers the pump. If there is no RPM signal it leaves the circuit open. You will find slight variations among control circuits. Ford circuits, for example, have an inertia switch that disables the pump in the event of an accident. GM's control circuit has a relay bypass that powers the pump through the oil pressure switch in case the fuel pump relay fails. Knowing how these systems work will help you size the circuit to their needs and also find problems quickly.

The ECU

An engine control unit (ECU) is a computer that electronically controls engine operation. It achieves this by processing input from sensors and adjusting engine functions based on the data. The simplest ECU's control only the quantity of fuel injected into each cylinder on each engine intake cycle. More advanced ECU's, found on most modern cars, also control the ignition timing, variable valve timing (VVT), the level of boost maintained by a turbocharger or supercharger, cold start functions, wide-open throttle (WOT) functions, adjusting for altitude and changes in heat/temperature, among other things. Originally these computers required very little amperage. Today their amperage draw is significantly higher because they must do more processing of more features.

ECU's determine the quantity of fuel, ignition timing and other parameters by monitoring the engine through sensors. These can include a Maximum Absolute Pressure (MAP) sensor—something like a vacuum gauge—throttle position sensor, air temperature sensor, oxygen sensor and many others. Often this monitoring and control is done using a control loop (such as a PID controller). Before ECU's, most engine parameters were fixed with the quantity of fuel per cylinder per engine intake cycle determined by a carburetor (a calibrated leak) or mechanical injector pump on diesel engines. All of these components are wired

This is an ECU for a F.A.S.T. fuel injection system. It is loaded with a variety of fuel maps as well as a baseline fuel map for starting the engine after installation. It is highly programmable and will deliver exceptional performance, depending which engine it is used on. It too has a high amperage system demand at maximum performance. A full installation of a F.A.S.T. EFI system is covered in Chapter 10.

This aftermarket system by MSD will control timing on a non-computer vehicle as well as an ECU-controlled vehicle. It will provide multiple sparks at low and moderate rpm and deliver a full +40,000 volts to the spark plug wires through its special coil. But don't forget to consider the additional amperage load it will add to your vehicle.

This GM/Delco ECU (computer) handles all the functions on an S-10 pickup. It has no distributor, but fires each cylinder on time based upon sensor inputs. Start adding accessories that draw a lot of amps and performance could go down unless you upgrade the system.

through the master harness by the manufacturer of the EFI system. That is why a modern EFI aftermarket system demands a 40-amp relay.

Function—A modern ECU uses a microprocessor just like your PC, which can process the inputs from engine sensors in real time. An electronic control unit contains hardware and software. The hardware consists of all the usual electronic components on a printed circuit board (PCB), ceramic substrate or a thin laminate substrate that you will find in your home computer. The main component on this circuit board is a Central Processing Unit (CPU). The software is stored in special memory, typically in EPROMs or flash memory so the unit can be re-programmed by uploading new code or replacing EPROMS. This is also referred to as an Engine Management System (EMS). See Chapter 10.

There are other systems the ECU controls on modern vehicles such as variable valve timing and turbocharger wastegates. They also may communicate with transmission control units or directly interface electronically controlled automatic transmissions and traction control systems. The Controller Area Network or CAN automotive network is often used to achieve communication between these devices. Modern ECU's also include features for cruise control, transmission control, anti-skid brake control, suspension control and anti-theft control. Every new job an ECU is assigned adds to the power it will consume and the demand on the electrical system.

As usually occurs with a technology shift,

computer-controlled engine management has replaced old failure modes with new ones. With advanced age, a failing ECU can cause seemingly random starting and drivability faults. For example, a vehicle may refuse to start when cranked with the starter motor, but may respond easily to a push start.

Electrolytic capacitors in the ECU are used to guard the unit from variations in voltage. When these start to fail, they no longer smooth the power supply to the microprocessor and the result can be sufficient line voltage fluctuation to reboot or restart the computer repeatedly while attempting to start the engine. An industry has evolved to refurbish ECU's with this and other types of failures related to age and use. So when you have a problem, get on the Internet and find a company

The cargo area of this Cadillac Escalade is crammed with 4,000 watts of amplifiers and big bass speakers. Driving this system requires a lot of amperage and a very good charging system. It has been fitted with two batteries and a high amperage alternator. The amplifiers are mounted two per side.

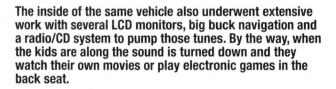

The inside of the same vehicle also underwent extensive work with several LCD monitors, big buck navigation and a radio/CD system to pump those tunes. By the way, when the kids are along the sound is turned down and they watch their own movies or play electronic games in the back seat.

Even the tight confines of a Mini can handle a high-dollar sound system and some extra speakers in the back as well as amplifiers under the seats. The Pioneer Tuner/CD player delivers quality sounds to the tight interior. You really do not need 1,000 watts of speakers to fill this little car with tunes. It actually presents little more than the stock sound system had demanded, so the owner has little to worry about. See Chapter 12 for a how-to installation.

that will refurbish your ECU. But first you should check the voltage drop of the ECU. (Running amperage tests on your ECU should be left to professionals. See Chapter 7 for more on ECUs.)

The Bottom Line

Adding aftermarket accessories and performance items will increase the load to your vehicle's electrical system, the question is by how much? You will need to account for them and adjust or upgrade your components to meet the extra draw. You audit all devices installed and write down the actual amperage and maximum draw numbers for each component and add them up. Aftermarket components usually list the draw in their instruction manuals or data sheets. Then compare the actual load to the rated load for your system. If the total maximum draw for all your individual circuits exceeds the maximum amperage available from your battery or alternator, you had better upgrade it. You should also be able to identify which new circuits draw the most amperage. This includes everything from off-road driving lights to 2,000-watt audio systems.

HOW TO USE A MULTIMETER

A multimeter is a an electrical meter that can perform several tests. This will be the most important electrical tool in your toolbox. When you have problems you will reach for this first. It can tell you AC voltage, DC voltage, AC amps, DC amps, resistance in ohms, and perform continuity testing.

To test for voltage, set the meter to V with the solid and dotted straight lines above it. Then touch the red probe to the positive side of the circuit and the black to the negative. Knowing which is which is easy at the battery, but can become difficult at other points of the vehicle. If you get them reversed the reading will be negative volts. Switch them and you will get a true reading.

Testing for resistance in a wire or circuit is really simple and informative. In the case of spark plug wires it is important to know that all 8 plug wires are close in total resistance. Since these are noise suppression (electrical feedback fighters) they will have a lot more resistance built into them than plain multi-strand wires, but that isn't the whole story. The most important thing in plug wires is that they are close in resistance. If one is really high, you know it is burned out and needs to be replaced. This is like compression testing cylinders. The key bit of info you are looking for is that they are all close, without any single cylinder more than 15% above or below the others.

If all we wanted to know was continuity (the voltage going in is the same as going out), we could move the dial one more stop clockwise and test it the same way. If it made a beep sound there would be continuity. This is very useful if you are looking to find both ends of a circuit under the dash for example. Just hold one probe on the end of the going-in wire and then touch the ends of what you suspect might be the coming-out wire. When you get the correct one, you will hear it beep.

You can test for how many amps are going through a wire or circuit. To do that, disconnect the wire either going in and/or going out of the circuit and move the dial all the way clockwise. Then with power on to the circuit and the connected components running, place the red probe on the live wire before it goes in and the black wire where the live wire was connected. This is like using the tester as an extension wire.

The reading will show the amperage draw by the circuit and/or component. If you are testing a 10-amp circuit and you are getting 15 amps, you will know something is wrong with the component (probably defective) or if it is a circuit it will be a ground wire in most cases. Fix the ground wire and test it again. If it isn't fixed you have something drawing amps in the circuit that shouldn't be there. It is probably an open wire contacting another circuit or component.

A multimeter is just a tool to help you locate problems; it doesn't fix things by itself. For that you need a troubleshooting strategy, which is outlined in Chapter 14.

The multimeter can perform several tests. Starting from the top, is the setting to test AC voltage (note wavy line), the next is to test DC voltage (note solid and dotted straight lines), 300-volt millivolt DC current, followed by resistance testing in ohms. After that there is continuity testing (with and without a sound signal), followed by amperage testing in AC current first and then DC current. These are all the tests needed, because you can do voltage drop with them as well as isolate broken wires and bad grounds.

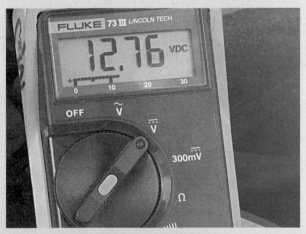

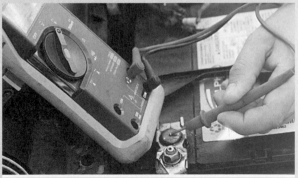

This battery is reading 12.76 volts with the engine off. This is a simple voltage test, with the red probe on the positive side and the black probe on the negative side. Note that both the red and black probes are going into the meter on the right side. These are the correct positions for voltage and continuity testing. Only amps use the other side of the meter and only for the red wire.

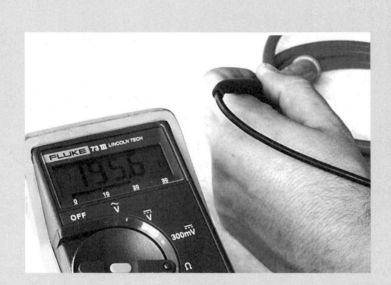

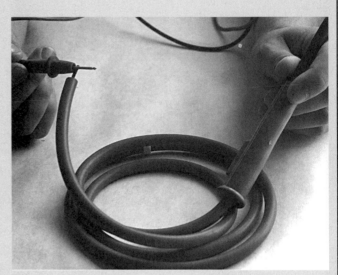

First look at the places the probes are placed. Touch either end of the wire. It doesn't matter which probe goes where, the result will be the same regardless. In this case the reading on the tester is 195.6 ohms. This MSD plug wire reads quite low for this type of noise-suppression wire.

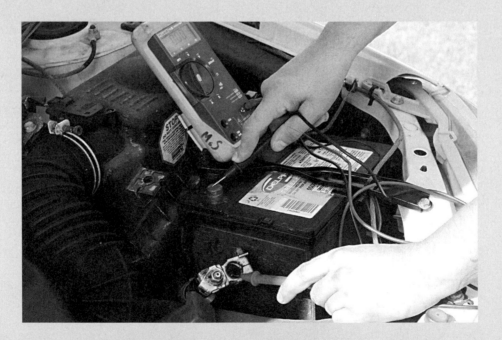

Here you can see how we disconnected the battery and placed the red probe on the positive battery cable and the black probe on the positive terminal. We have the engine running so we are reading the amperage returning to the battery from the alternator.

Dead-ending wires or terminating them properly is just as important as connecting them. In this photo a skilled tech from International Motor Sports is terminating a group of extra wires from an electrical system that will remain in the vehicle due to class rules for this race car. He cuts all the wires cleanly and then uses shrink tubing to hold them in place and leave it a bit long so he can put in that extra tubing and squeeze it together to seal the wires from moisture and dirt.

Chapter 4

Making Electrical Connections

Most of us have developed our automotive skills through trial and error, a time-honored method of learning. After a few of these trials are successful, you will be considered a reliable and trusted member of your local brotherhood of car geeks and/or hot rodders.

A good example is the basic knowledge you probably already have picked up about fasteners. You have learned a great deal about bolts, nuts, washers, screws, rivets, cotter pins, lock clips and hundreds of other similar methods of attaching things together, just from working on cars. But electrical systems require their own specialized set of connectors and connecting methods. The two main methods are soldering or using crimp connectors.

You can elect to solder every wire and connector in your hot rod and I must admit it would look really trick if it was all soldered carefully with shrink tubing over every single joint. There are advantages to using solder, but to do every connection would require a lot of work. The quickest way is to use crimp-style connectors, which you'll find in most vehicles, both OEM and custom.

Crimp Connectors

A good knowledge of the many types of crimp terminals and connectors is necessary for you to take advantage of them when building a car or doing wiring repairs. There are butt connectors, snap connectors, eye wire ends, spade wire ends, slotted wire ends, two-into-one connectors, water

resistant vs. regular—most of them available on the web or at your nearest auto store. You'll also find many types and sizes of shrink tubing, wire, junction blocks, fuses, fusible links, fuse panels, etc. A great supplier is Painless Performance Products, which specializes in aftermarket automotive applications, from classic restorations and muscle cars to racers and modern street rods.

There are basically five major types of wire terminals/connectors used on vehicles: flat quick disconnects, butt, ring, bullet and the U-spade, which is designed to slide in under a washer and nut on an electrical stud or screw. Flat quick-disconnect connectors allow you to stick in a female connector as a quick disconnect for a circuit. Bullet connectors do the same, but use a different shape. Butt connectors are used to connect two wires end-to-end. Ring connectors provide a ring to go around a bolt or stud terminal. They are easier to integrate into a wire harness than wider quick disconnects. U-spades are like a ring connector, but have one open side. I stay away from these except where they are critical in places like instruments or other electrical service points.

There are also many variations to each. Some are water-resistant or fully waterproof, some connect more than one or two wires, some connect two wires without cutting them, etc. You need to do your homework before you start using crimp connectors.

On many automotive circuits you will find two, three and

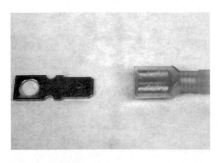

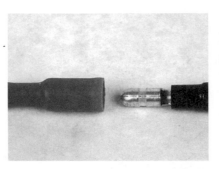

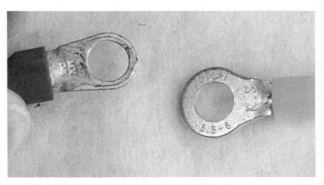

These are flat quick-disconnect connectors. The 1/4" flat female is on the left and the male is at right. They are made to be pushed on and pulled off quickly. (Note: The male end has a hole for attaching it to an instrument, but they come in regular male styles also.) The bullet connectors (male and female) are excellent for joining two wires end to end, and having an easy quick disconnect.

Ring (shown) and U-spade connectors are made for a specific size stud and will have a number that corresponds to stud size: 4, 6, 8, 10, 1/4", 5/16", 3/8" or 1/2". The larger the stud size, the more current that connector can handle.

A ring connector is mounted on a stud or fastened with a screw. They use the fasteners as conductors to keep the circuit going and to keep them from falling off and causing other problems. At right are flat quick-disconnect connectors made to be slipped onto a terminal connector with walls to contain and lock in the lugs. This connector is for a 2-wire hookup on a GM HEI distributor.

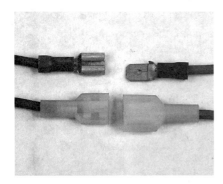

The difference between standard insulated connectors and weatherized connectors is clear at left. You should use a ring connector (right photo) for studs because it is stronger and less likely to fall off, even if the stud nut loosens. If you use a "U-spade" connector on a stud you may find that the lugs spread as you tighten the stud. Also, if the nut or screw holding a U-spade connector becomes loose, the connector may slip off, causing a short.

more connectors, making troubleshooting and repair easier. For example, a two-wire GM HEI distributor connector might also be the same connector used for the A/C clutch as well as the alternator. My preference is for quick disconnects and ring connectors. A quick-disconnect installation makes a good electrical connection and is mechanically sound. In the photos above, the standard connector and the water-resistant insulated connector are shown to make the differences visible. The fully weatherized insulated connectors should also have shrink tubing over the ends where the wire comes out. Then they would be waterproof, not just water resistant.

For connecting two wires end-to-end, the standard butt splice is a great connection. They are all available color-coded to match the wire size and the correct position on the crimping tool. A butt splice that needs no tools to crimp is also available. Look for the correct color code for your wire size. These connectors fasten by simply twisting each end to the second position after they are inserted. They are great for emergency repairs on the road, but I prefer a well-crimped connector for a longer-lasting connection.

Color-Coding—Crimp connectors are color-coded to indicate the wire gauge size. Red is 16-to-22 gauge; blue is 14-to-16 gauge and yellow is 10-to-12 gauge. Anything bigger (starting at 8-gauge wire) will need to be crimped with special heavy-duty industrial electrical tools.

Some people use non-insulated crimp connectors and then put heat-shrink tubing insulation over the wire and connector to help support it. This is an ideal way to go if you need to make the connection strong and weather- or contact-proof. Since shrink tubing comes in many diameters, colors and types (shrink factors such as 2 to 1 and 3 to 1) plus

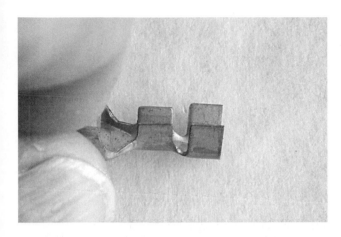

The best strain relief a connector can have is two extra metal crimps at the base going right to and on the edge of the insulator. This is important as this style crimps the wire for a strong electrical connection and also crimps the wire insulation for extra support.

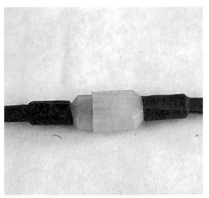

Here is a water-resistant spade quick disconnect. The terminal ends are covered with shrink tubing (3-to-1 shrink plus self-sealing glue) that will repel most water problems. But to make it waterproof you can seal it with clear silicone where the two spade halves meet.

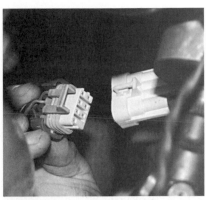

This photo shows a "Weatherseal" connector. These actually have an O-ring on each pin wire so water cannot enter the connector down a wire and the entire connector is also weather sealed once it is snapped into place. A very cool but expensive connector system.

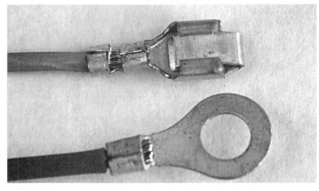

The ring connector on the bottom has only the crimped wire for strain relief and the wire will eventually break due to movement and vibration. The female spade connector on top uses the wire insulation as strain relief and will last a longer time before failing. These type tend to be a little more expensive, but the extra protection is worth it.

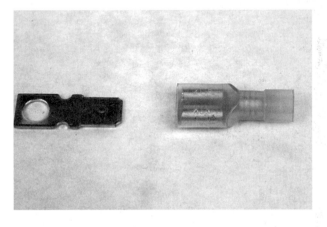

This is a quick disconnect connector for using on an instrument post or other threaded post where you want a quick dismount without having to remove a nut and washer. The male is on the left and the female on the right. The female end is butt-spliced onto the power or ground wire of the component.

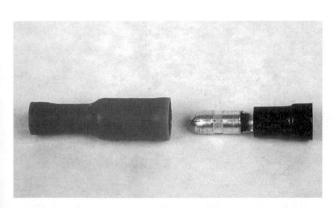

This is a typical female (left) and male (right) round connector that can be butt-crimped to wires so they can be disconnected easily for removal or circuit testing. They come in different sizes based upon wire size.

features like glue inside, you need a knowledgeable supplier to help you make the right decision.

Quick Disconnects—OEM manufacturers use quick disconnects in many places on an auto or truck. If you are making your own harness you may need to reuse some of them. If that is the case, remember to leave long pigtails on the wires you cut so you can splice in the new ones. You can look for these on the brake pedal switch, emergency brake light switch, instruments, grounds and other locations throughout an OEM wiring system. Please keep in mind that these quick disconnects may be made with as many as 8 to 12 wires going into each side of the connector. A typical example is the steering column.

The steering column houses quite a few electronics in today's modern vehicles, so they are typically jammed with wires. They are for the ignition switch, turn signals, headlights, high beam switch, windshield wiper functions, and on some

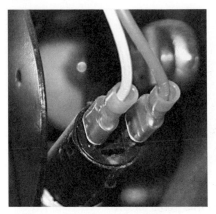

Typical connections for quick disconnects are such as this brake light switch. Both connectors are insulated and slip over a male spades built into the switch. Just one side is insulated as they cover the male end completely.

This is the steering column connector for a 1974 F-body Chevrolet. The column was salvaged complete with the key, wiring and connector for installation into a 1956 Ford Pro Street truck. The connector half on the right came from Painless Performance. It matched perfectly to the factory side of the connector on the left. Although it is a 10-pin connector, we just used 8. We matched the colors to the factory ones, leaving out the high/low dimmer switch wires for the column because we used the original floor dimmer switch instead.

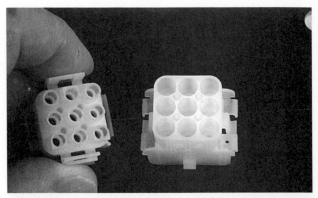

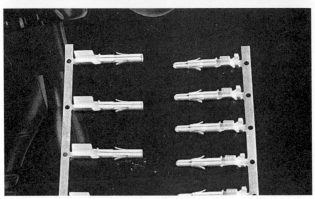

The white connectors in the top photo were used to make the dash insert in the '56 Ford pickup an easy slip-in and slip-out operation. They can only go together in one position so it is important to get your wiring for each side to match the first time around. Removing the pins to move them to another position is very difficult so check position three times before inserting a wire and pin. In the bottom photo you can see the pins as they were purchased with the connector. The male pins are on the right and female on the left. Note the arrow-like locking mechanism to keep the pins from coming out. These are why it is tough to move them if they are in the wrong location.

This aftermarket two-wire connector for a simple circuit is a very common connector on automotive vehicles. These are generally made to fit into OEM harnesses and components like alternators, headlights, and a host of other things. But, you may find some are generic and you may decide what they connect to and use the connectors close to components to make removal easier. Note that the female pins that go inside have a double metal wrap. One is to grab the insulation and the closest one is for the wire. You must crimp both of them to make sure the connector is going to stay together and not cause intermittent connections.

vehicles, electronic shifting. Because of this there are many circuits going through the column. Generally there is a main connector on the column that allows you to test circuits easily without taking things apart.

We chose to use a 9-pin connector (see photo above right) for a quick disconnect to the dash instrument insert on the '56 Ford pickup shown in these photos. The male and female pins are specially designed for this type of connector. They are small and require a gentle touch, but they do a good job. These connectors are designed so they can only go on one way, assuring that the wires stay in their proper position relative to the other half of the connector. You must match wires correctly the first time or you will have to carefully remove the pins and reinsert them correctly. (A real pain—you have been warned.)

The round pins (male and female) that go inside the 9-pin connector have the same type of double

metal wrap as discussed earlier. The small one grabs the insulation and the large one the wire. Be careful during assembly. The female uses a large crimp area to crimp both the insulation and wire. The things sticking out like the barbs of an arrow are the locking device of the pin. To remove a pin from a connector you must hold both in and pull the pin out of the connector at the same time. That is a difficult task so get it right the first time. Obviously, the reason this has been mentioned several times is that we have first-hand experience having to remove a couple of them and it is a real pain.

Solder or Crimp?

This question always causes heated debate between good mechanics as well as hot rodders. Old-school guys will recommend soldering and the new-school guys will tend toward crimping. For

One of the main disadvantages of soldering is that solder wicks into the wire and stiffens it. If the wire is then flexed it is likely to break at the joint. Because of this, make sure you limit its application to things that will not vibrate or move. Here you can see that the joint has not stayed together (because it was made by simply pushing the two wires together) and the heat and solder has wicked way up the two wires' ends. It was stiff a full inch up on both sides. This joint is weak and the wire will break if used as is.

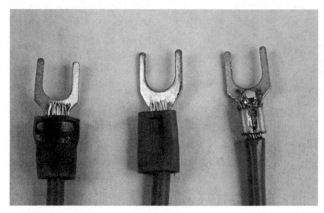

All three of these wire ends are poorly made. The first was crimped with the wrong tool and too much wire is coming out of the top, the second has too much wire going though it and the last has been soldered after it was crimped. That lump of solder at the top end will not allow this connection to stay flat and make full contact.

many electrical joints there is no choice but to solder (e.g. joints internal to an electronic device such as an ECU/computer or amplifier). Both crimp and solder have their uses when building or repairing a street-driven stocker, a hot rod, street rod, muscle car or race car. If solder is done correctly, it will make a good connection.

Solder Advantages—A properly soldered electrical connection has its virtues. Solder reinforces the mechanical connection of joined wires and helps to prevent their inadvertent separation. Solder is an excellent conductor, and causes minimal voltage drop. Solder allows for visual inspection of the exterior of the connection. Solder has an almost universal application, regardless of differences in wire gauge, component location, available space, and component function. Solder is also readily available from any hardware or home supply store.

Disadvantages—Solder is hot! Even with a heat sink, solder may seep into areas beyond the intended joint or may cause damage to adjacent wires and components. Solder can trap moisture that eventually may corrode a connection from within. Too much solder may reduce the electrical conductivity of the stranded wires. Too little solder may cause the joint to fail. Solder does not permit visual inspection of the interior of the joint.

A soldered connection will be less flexible than the original stranded wire, and may weaken over time due to vibration. A soldered connection must be insulated by heat shrink, by electrical tape or by another insulator. A soldered connection is

permanent and cannot conveniently be disconnected. And many of us lack the knowledge or skill to solder properly.

Sometimes, we strike a compromise that combines the worst features of both. Most crimp terminals are designed to be crimped, not soldered. If the crimp was done poorly, solder won't save it. And if the crimp was done properly, solder is unnecessary. In fact, soldering a crimped terminal may weaken the mechanical connection, may reduce electrical conductivity and may damage the terminal. As a general rule, you should not solder a crimp terminal.

Mike Priore, the owner of the 1972 Barracuda we used in several chapters of this book, offered his opinion on whether or not to solder. Mike is a line mechanic at a Honda dealership so his years of experience using crimp connectors is valuable. He states:

"After 14 years on the line at new car dealerships and attending training schools by the OEM's, there appears to be no reason to take out the soldering iron or gun anymore. Most good mechanics and technical people will tell you that crimped joints are as good as or better than soldered. But only if you have the correct crimping tool and know how to use it. You also need to use the correct size connectors for the wire size. Crimping a tiny wire into a connector intended for a large size wire will not work. Even worse is cutting bits off the wire to get a large wire into a connector that is too small. However almost all electrical joints in modern vehicles are crimped and that isn't because the OEM's can't afford the solder, so there must be a good reason."

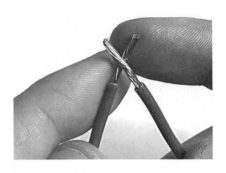

Here we are sweat-soldering the industrial-size terminal end to the battery ground wire where it will be attached to the frame on the '56 Ford Pro Street, just 2 feet from the battery. This is just like soldering copper tubing, get everything very clean with a wire brush or sandpaper.

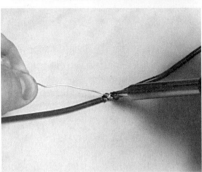

Clockwise from top left: The first task is to strip the wires, then twist them so they support each other from being pulled apart. Next, heat the wire and solder it. Use the sweat solder method by getting the wire hot first and then when you apply the solder it will be sucked into the wire strands.

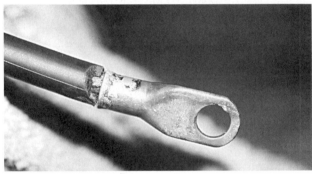

Here is the completed soldered terminal. You can tell it was hot, but you can also see that the heat never went above the top of the terminal end, because the insulation isn't melted.

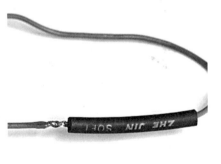

Ideally, soldered joints should be mechanically supported, so by twisting the two ends over each other the joint becomes self-supporting and the solder just completes the circuit properly. An added piece of heat shrink tubing around a joint helps take strain off the solder joint itself.

much you burn the insulation. With the right temperature the solder will suck into the wires. Then let it cool without moving it for a full 60 seconds. The result is a good joint. If you get it wrong you will either expose the wire too far back with burned insulation or you will have a dry joint.

Switches

Switches can be activated by either vacuum, heat, cold, pressure or manually with your finger. You should always try to work with new switches, whether you are restoring, repairing, or building from scratch. For example, on one project we needed a Chevrolet light switch for late '60s-through-'80s GM F-body cars. This switch can be used on many vehicles, but it is also great hot rod material. We adapted it to our project 1956 Ford pickup shown throughout this book. We ordered the Painless Performance Kit 16 circuit/GM kit which gave us the connector for the switch and all the correct wires. We also ordered a new switch

Soldering Method

First you have to strip the ends of each pair of wires and either twist them together or force them together by pushing one end of wire into the other end. I prefer the twist method as each will become self-supporting to the other wire when they are twisted completely.

Finally you can solder them using thin rosin core non-lead wire. The trick is to heat the wire quite hot before you touch the solder to it, but not so

The terminal installation was completed when the shrink tubing was heated and it conformed to the terminal and insulation of the cable.

Thermal switches can close an open circuit once they reach a preset temperature. These would go into the wet-side of the intake manifold. But they can also open a vacuum line, allowing vacuum to go to a component once the engine temperature reached a predetermined level.

When working on older cars and light trucks, it is always best to buy new switches if they are available. Switches have a cycle life and wear out just like an engine, transmission, etc. Why risk potential breakdowns at the side of the highway when the solution is simple as you are building or rewiring your project vehicle?

Pressure Switches—Pressure switches can be used to keep electrical circuits open when there is pressure in the system, or close them when there is no pressure. Both of these switches are calibrated to activate at 50 psi. The one on the left with two flats close the circuit so it can operate when there is pressure and open it when there is not 50 psi. The one on the right with one flat is calibrated to stay open at 50 psi and close below that. These are for special service and functions, but keep them in mind when you are creating new systems.

Safety Switches—Safety switches are just as important as a light switch. I lost a friend at a race track as he was walking through the pits. Someone had not wired a neutral/park safety switch into their starter circuit. The car took off and pinned him between a pickup and the race car. Never wire a starter circuit without the neutral/park safety switch.

I use the pressure switch at the left to open the fuel pump circuit when there isn't engine oil pressure. That keeps the pump from continuing to work and causing a potential fire if the engine blows up in a race or on the street. You could also use it to shut the engine off if the engine loses oil pressure, or even if the fuel flow drops below a specified pressure.

because working with old connectors and switches should be avoided except when there is no alternative. All we had to do is plug the wires into the connector and snap it onto the switch right before mounting it into the dash.

Thermal Switches—Thermal switches are sensors that close or open a circuit when they reach a preset temperature. Typically they are an integral part of an EFI system and an automatic electric fan assembly. Thermal switches and sensors can have different connectors. Some take typical female flat quick-disconnect connectors. When you are talking about an EFI system, Weather Pack connectors are used to keep them watertight. These are what send the engine temperature information to the computer (ECU—electronic control unit). This information is critical to the operation of an EFI system.

Switches designed to shut off all electrical power from the battery to the vehicle are available in several forms. The most common is the outside switch with a twist handle. There are others that use a solenoid that opens or closes the main power circuit from the battery to the vehicle. As solenoids (a type of relay) they do not require a lot of current

Both of these master electrical cut-off solenoids are designed to be inside the vehicle, under lock and key. Pressing a single button either opens them or closes them based upon their present position. I also like these devices for street cars as they will defeat the average thief, or make him spend too much time in the car trying to find the correct wires.

This is a simple micro switch that can be used for many applications. The roller wheel is contacted by a lever that moves (throttle, parking brake, brake pedal, etc.). This switch is mounted to the carb in a position where the throttle linkage will close the circuit at wide-open throttle (WOT).

In this photo you see the safety switch below the nut of the shifter handle pivot. It will only close when the gearshift is in neutral or park. On a manual shift vehicle, wire in a safety switch so it will not close the switch unless the shifter is in neutral and/or the clutch is pushed all the way down. The wires from this switch come from the key and go to the positive post on the starter relay.

to open or close a circuit, but they can handle all +70 amps of power that is in an automotive electrical system. On a street rod, being able to press a button and open all electrical circuits in the vehicle is a good anti-theft device as well as a safety device

Micro Switch—A micro switch can be used anywhere a motion needs to be used to close a circuit, such as with a nitrous system. You just make sure the roller wheel is in contact with the throttle lever, so when it is wide open (or sooner, if you decide), it will close the circuit for the nitrous and start the spray. Micro switches can be useful in many circuits. For example, they can be used to activate the door locks any time you have the throttle above idle. They could be used to make sure the A/C will not kick in unless the engine is above a certain rpm, or many other uses when closing a circuit might be controlled by the throttle. Micro switches can also be used as a brake switch, an emergency brake switch, etc.

Rocker Switches—These are used in place of toggle switches mainly on race cars to prevent accidental activation. Using rocker switches that light up when the circuit is on can eliminate issues such as running off-road lights in the city, or running a manual switch for an electric cooling fan for the radiator. When the light is on, the circuit is active—a simple, quick indicator that a circuit is on.

Switch Panels—Switch panels are very professional looking and useful. I love them on any hot rod because of their looks as well as functionality. On race cars they are essential because they make it easier for the driver to operate controls for many systems. Fuel pump(s), starter, nitrous system, ignition box(es), driving lights, main lights, running/marker lights, etc. The more control of the vehicle the driver has within easy reach, the better he can perform.

Miscellaneous Bits

Although some of the following items are not connectors, they have been included in this chapter because they will facilitate many jobs when wiring a vehicle and setting up circuits.

Lighted rocker switches are useful because they let you know that a specific circuit is operational and active. This can be a distraction at night if you have a lot of them, or if they are placed too close to the driver's line of sight. They can also be a distraction when several lit rocker switches are placed in a straight line, unless they are different colors. (Different-colored rockers are available: red, orange, blue, green and yellow.)

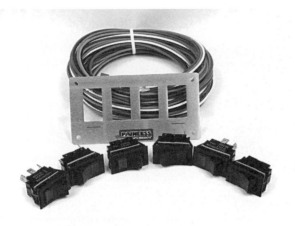

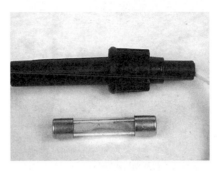

Painless Performance makes a simple 4-switch panel that works with their 16-circuit kit. They send you 6 switches so you can pick the 4 you need for your particular installation. They supply the wiring and terminal ends, but it is not prewired. This will do just fine for most hot rods. You can add driving lights, make your electric fuel pump operate with it, use it for a starter switch and many other things.

Making sure a circuit has protection is a critical factor when wiring a new circuit or repairing an old one. Inline fuses are simply spliced into a power wire of a circuit, giving that circuit protection for the amperage it requires. These two types (blade and glass) will be adequate for most wiring needs. It is also possible to wire a complete vehicle using inline fuses without using a standard fuse box, but you will have fuses all over the place and need a very good wiring diagram to find them if needed. It is much better to have the fuses in one place.

Painless also makes a prewired 8-switch panel. It comes with names for many functions, but you can change them if you desire. Having everything prewired makes it easy to install. One wire for ground and another for power to the switches and switch lights, then run the connections to the things you want to activate with the switches. Once you mount the panel box in the dash or below it, within easy reach of the driver, it is only about an hour or two to complete the installation.

These two photos are typical of the types of junction blocks used in vehicles. The first small one has room for one power-in on one post and power-out using several wires on the other. The power distribution block at right was taken off of a 1976 Chevy pickup. The power-in is on the bottom side and there are 4 power-outs. A distribution block is a convenient place to test for power in circuits. If there is no power here, the rest of the circuit will not have any either.

Inline Fuses—If you have to go into the vehicle and add a circuit (driving lights, audio equipment, etc.) it may be easier to draw power from an unfused source and use an inline fuse to protect the new circuit. (Use whichever you have installed in the vehicle to protect other circuits; either blade or glass type.)

Junction Blocks or Buss—A junction block is an efficient method to run power to circuits. One wire feeds power into the block, where it is dispersed via several posts and wires out to various components.

Junction blocks come in many sizes and with many connecting posts. Some of these mix connector posts sized for high current and others for low current on the same block. These are best located near the main power lead from the starter circuit or positive side of the battery. Just make sure you are not trying to run too much current through a small wire. Use fuses to prevent this from happening.

Adapter Posts—Adapter posts will allow you to convert a GM-style, side-post battery to one that will accept normal battery cables. They will be vertical, but they will work. These can be useful when when relocating the battery to somewhere other than the engine compartment, because they will convert cables going straight down without having to bend them.

Grommets—Always use a grommet when you are

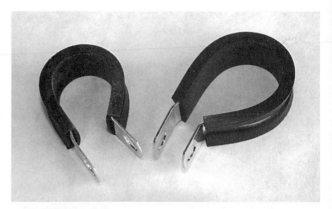

Some people prefer big battery terminal cables over the side-post connectors typically found on a GM product. These battery posts screw into the side-mount connection on any GM-style battery and allow the use of a standard-type battery connector. When mounting a battery in the rear of a vehicle, it is a good idea to buy a side-mount battery and then use these adapter posts so you can have the wires going straight down without having to bend them. This is a real asset in this type of installation.

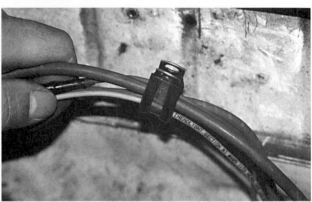

This simple insulated Adel clamp is essential when wiring anything in cars, planes, boats, racing vehicles and many other types of applications. They hold things tight, keeping them from moving and abrading. They also keep things up and out of the way so what they are holding is safe and looks professional.

Grommets come in many sizes so there is always one that will fit your needs. Try to get your wires or harness to go through the grommet with a tight fit. If you have sent several individual wires through, it is best to rub a bit of silicone over the grommet on the inside under the dash so water doesn't get in. By doing it on the inside you will not have a chunk of silicone visible where people will look at your vehicle and its build quality.

running a wire through a metal panel, or any other type of panel that could abrade it. They are not expensive and they can keep you from having major problems, the kinds of problems that can burn your vehicle to the ground and the building it is parked in as well. Without grommets, the question isn't "if" you will have a short to ground, but "when."

Adel Clamps—Technically, these are insulated metal clamps to hold tubes, hoses and wires in place in any area subject to heavy vibration, like in race cars or aircraft. They are used around fuel filters, electric fuel pumps, fuel and oil pipes and hoses, and also around electrical harnesses. Always keep a supply of them around when you are building or repairing electrical systems, they come in many sizes as small as 1/4" to as large as 12" from auto parts and specialty stores.

Adel clamps keep things from chaffing, keep things out of the way and protected from damage (depending on where you mount them). This clamp will keep them all together and protected from harm on the firewall.

Circuit Interrupters—These devices control the flashing of directional signals and also of the roadside emergency signals. Most automotive wiring systems require two. They can also be wired

into doing other jobs such as spraying water or sticky stuff on tires before a burnout, spraying water on the radiator of a vehicle that runs hot going up hills, when towing or in slow traffic. They are simple circuit interrupters and only creativity limits their application.

Split Wrap—When you are building wiring harnesses you will need some sort of split wrap to protect it. There are versions that are hard plastic and can take a lot of abuse, but it is not as flexible and limits where it can be mounted. There is also a softer version made by Painless that is also highly protective and more flexible. Using either of these wraps will not protect the wires inside from hard hits such as when crashing into walls with other vehicles or rubbing paint where the harness may be located, but for everything else they can provide a high level of protection. I have also seen some builders use conduit for any wires running from front to back. That might be a really good idea for an off-road vehicle, but only if it is kept inside the vehicle or frame.

The turn signals and emergency flashers are controlled by these circuit interrupters, generally called flashers. Both circuits use the same flasher so one spare would be all I would carry in a vehicle emergency kit. Some are tough to find because they are not mounted on the fuse panel and might be mounted somewhere under the dash. These are usually covered by a panel so they can be hard to find.

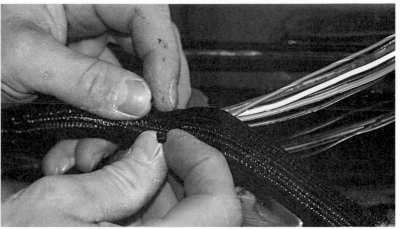

Steve Agnello is sliding wires into a 7/16" size Power Wrap. You can see that he has already installed one wire tie to hold the Power Wrap closed below where he is working.

Once the wires are installed in the wrap, you should secure the line with clamps or a tie.

We are shrinking tubing over a group of wires we will no longer be using so they would be protected from shorting whether they were dead or hot. You can cut each wire different lengths so they are staggered and cannot touch, or you can also cut them the same length. But, in either case you must make sure they are all waterproofed and tightly bound with the open ends covered with heat shrink.

Heat-Shrink Tubing—Heat-shrink tubing is a miracle product for electrcial wiring. I buy it in a 3 to 1 shrink ratio with glue insides so I'm sealing and protecting the terminal end connections. We made five ground straps for the '56 Ford Pro Street pickup to make sure the engine, chassis, body and bed were all well grounded. They are all a #10 wire with crimped eyelet connectors and heat-shrink tubing, for supporting the connector. The heat-shrink tubing was already on the wire before we made the crimp. It will shrink all the way down to become a tight fit, protecting the connector and providing a professional look.

Chapter 5
How to Read Wiring Diagrams

Symbols you need to know to read diagrams.

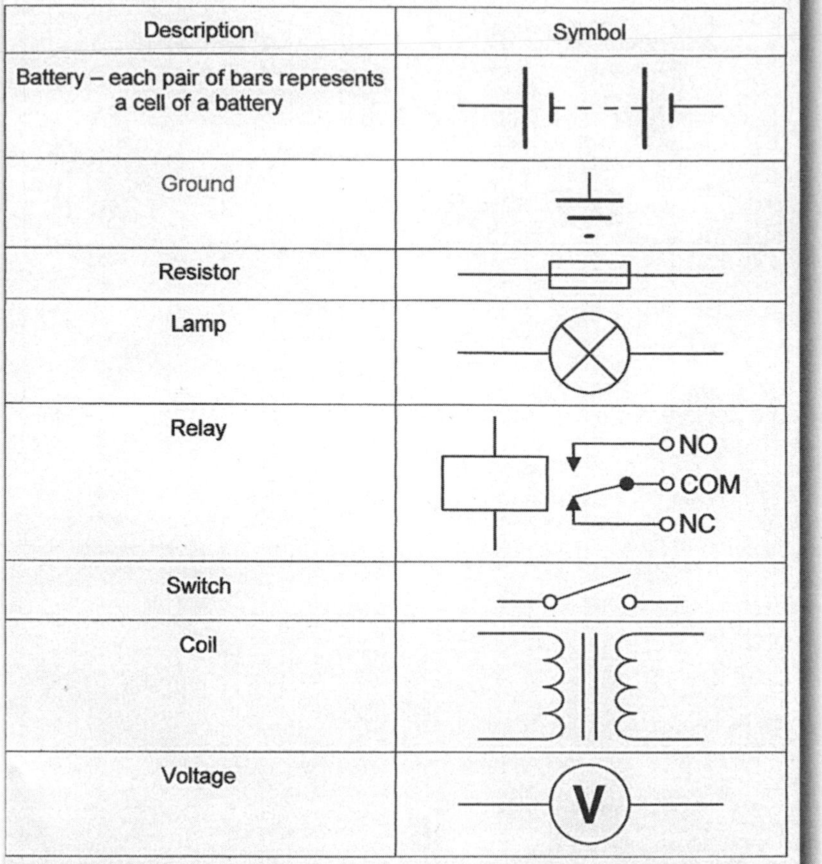

Description	Symbol
Battery – each pair of bars represents a cell of a battery	
Ground	
Resistor	
Lamp	
Relay	NO / COM / NC
Switch	
Coil	
Voltage	V

This chart presents the basic eight symbols you need to know. Starting from the top they are: battery, ground, resistor, lamp, relay, switch, coil and voltage. On a couple of these there are common variations; for example, the lamp could also be called a bulb and they would generally reflect the number of filaments inside each. The relay could have the switch part inside the box along with the symbol for a coil since these are the elements inside a relay. Switches could have many lines going from one side to the other and voltage will usually have a number included to tell you how much voltage is going into each component and/or circuit. Sure, an automobile is 12 volts, but some circuits may show 1.5 volts or even less when they are normal.

Wiring diagrams or schematics might look intimidating, but they are just roadmaps of wiring circuits. If you can read a roadmap, you will have no trouble with them.

If you buy a complete set of wiring diagrams for your project or pet vehicle, they should contain at least the following diagrams—1) dash wiring, 2) interior wiring, 3) exterior wiring, 4) engine wiring and any accessories that were options for your make, model and year vehicle. I wouldn't accept just one big diagram for everything in the vehicle unless it has been rewired using a non-stock kit, or wired after building a custom vehicle. Then I would want to know which wires were terminated, which do not go to what the master schematic indicates or the wires themselves say they are to be used for. Of course, this should be provided by the builder if you are buying something expensive.

The first thing to learn are the different symbols. The symbols we are presenting in this chapter are standard in the industry, but some companies use slightly different ones for propriety products. Yet, they always name them even if things look different. In the diagram on page 43, from Fuel Air Spark Technology (F.A.S.T.), you will see that their symbol for a "ground" (a circle) isn't the same as the ones we usually see.

As you are looking at a diagram, keep in mind which circuit your problem comes from and even if you are mixed up a bit, you will get through it. For example, what if the problem is the brake lights and you look at the lighting circuit. You might not find the brake light wiring in the lighting circuit, but since you know these bulbs are part of the tail lamp housing you will find the right bulb and can trace the wires back to the inside of the vehicle. Since you are looking at a brake light problem, you need to understand where the grounds are, if the bulbs sockets are self-grounding and where the actual brake light switch is located. (Usually attached to the brake pedal.)

Circles in a diagram are differentiated by a symbol within the circle (a motor, voltage, bulbs and lamps or even grounds). It may be spiral wire filaments or an X for bulbs, a voltage number (12V) showing how much voltage should be in the wire, an amperage number (15A) for a given wire and a callout if it is a ground.

Circuit protection is also important in a wiring diagram as it can tell you how many fuses, fusible links or circuit breakers are in the system and their values are usually shown

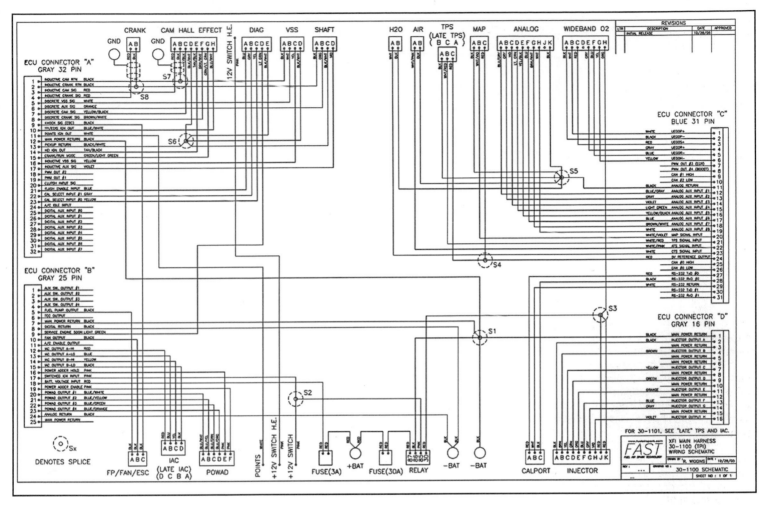

This wiring schematic or diagram is for the F.A.S.T. electronic fuel injection system. It looks really complicated and the fact that it is an electronic system might deter you from trying to understand it. But it isn't complicated if you look at it bit by bit. For example, all the wires on the right and left sides are simply the wires for the 4 different connectors that go into the EFI management box. The wires are listed by color and pin numbers for each of the 4 different connectors. If you follow the wires back to where they started you will see that all the ones on the top go to the sensors that collect information/data about the air temperature, engine temperature, throttle position, engine load, etc. The ones at the bottom are for things like injectors, relays, circuit protection, data port (where you can plug in your computer), switches and electric-fan control. Then, if you look at the entire schematic you will see that everything in the middle is just wires going from one place to another. For an installation of the F.A.S.T. system, see Chapter 10, page 141.

as a callout. 20A on a fuse would make it a 20 amp fuse, etc. Know these symbols because they are important.

Coils

Coils are a unique part of the electrical components in a circuit and it is important to identify them in a diagram. They either make something move or they change the voltage in a circuit. A coil is used in every relay in the vehicle. It is what allows a relay to close a higher amperage circuit with less voltage. It is also a safety device as it can reduce the chance of sparks from a high-power circuit when it goes from open to closed. Typically these are used throughout the vehicle because a relay lowers wire size/weight/cost

throughout the vehicle and generally extends the life of switches.

A coil can also actuate solenoids by delivering power to the magnetic field that makes an actuator move in and out. An electromagnet is the basis for relays to work; it is nothing more than a coil wrapped around a metal bar (or anything else such as those things that can pick up cars in junk yards). An electromagnet circuit can be turned off and on with a simple switch and relay circuit. Both of these types of coils are diagramed the same, but they will have different components because the coil inside a solenoid and the one in a relay look different.

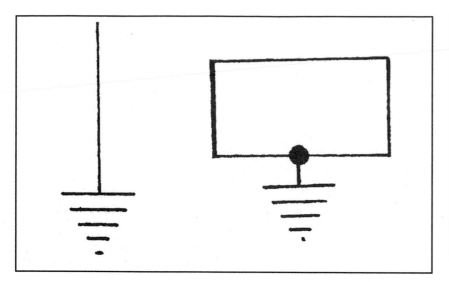

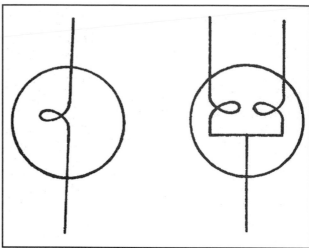

When it comes to grounds they are usually the upside-down Christmas-tree type, but in Europe they might look like a rake as well. In all instances they are critical to your understanding the diagram as well as knowing where to look for ground faults. If a ground has something stuck to it that is round, square or rectangular, it indicates that a component is "self-grounded." These components could be relays, horn(s), electric fans, solenoid(s), amplifier(s) and other components such as the starter. If the component suddenly quits working you would automatically do a visual inspection, check the fuse and run a test ground wire from the problem component to a good ground and see if it works. This is a quick test to remember on self-grounded components as it just takes seconds.

Bulbs/lamps are usually shown as circles with either an X or spiral wires for filaments. If an empty circle is to be used as anything else it will have a call out. If the circle has a number such as 12.75V it is showing you that the circuit should have that many volts. If it has 16.5A, it is telling you it should have 16.5 amps if the circuit is healthy.

Schematic Formatting

A wiring diagram must display the beginning and end of the wire or circuit, but of course that does not indicate the true length. You'll have to adapt the diagram visually to your vehicle, drawing or imagining where the wires would run under the floorboard, into the firewall, etc. To get the schematic on one sheet of paper, some formatting rules have been adopted. The first is that wires are drawn only in straight lines and they go from start to finish unless some form of bulkhead connector or power distribution buss gets in the way. These lines can take right and oblique angles to others, but they cannot curve. Unless they are connected or spliced where lines cross, the straight line must remain uninterrupted.

Let's look at the dash wiring diagram from a 1971 Barracuda with the 4-bbl engine and a 4-speed

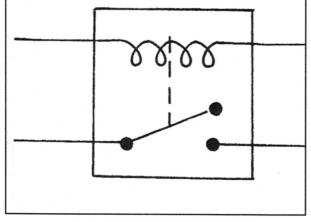

This is the most graphic representation of a relay because it shows the switch as well as the coil. That coil usually is for another, higher amperage circuit to engage.

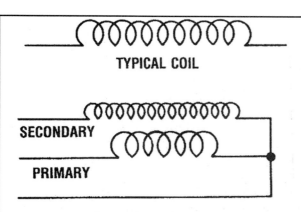

TYPICAL COIL

SECONDARY

PRIMARY

A coil that changes electrical voltage in a circuit (the ignition coil) is different. Its purpose is to raise the voltage from 12 to more than 25,000 so a spark plug can make the electricity jump from the center electrode to the wire part of the plug that is grounded to the head. This aftermarket coil is actually two coils in one. The outside coil has windings that are typical for any other coil in the 12-volt automotive system, but they are wrapped around another field and coil that has very fine wires. These fine wires can change the voltage from 12 to more than 25,000 volts. But to deliver this spark power to the plug requires that the field be suddenly opened so the high-voltage field collapses and that power runs through the plug wire to the plug at an incredible speed and power. Without this coil system, 12 volts could never generate enough power to jump the distance from the center electrode to the wire part of the plug. (Compression pressures and wet fuel would simply stop the spark before it could start.)

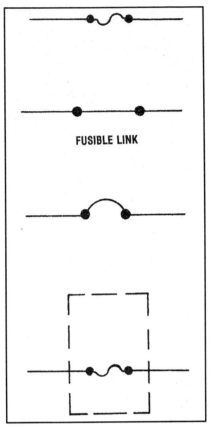

FUSIBLE LINK

The first symbol is a typical fuse, the next two are fusible links and the bottom symbol is for a circuit breaker that usually resets automatically for automotive use.

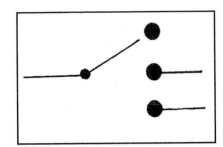 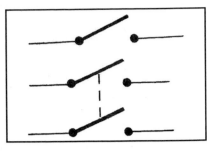

These are three different representations of the common switch. Every switch is nothing more than a circuit interrupter. It stops the flow of electricity or allows the flow depending upon its position. Open is when it stops flow and closed is when it allows flow. The first diagram is a single-pole, triple-throw switch by itself. This is to show you the difference between throws and poles. In this switch there is a single input of power (throw), yet it redistributes that power to three individual wires/circuits or components (poles). The other diagram shows a single-throw, single-pole switch, which is the most common type in a vehicle. It takes power from one side and sends it to the other when it is closed. The last is a double-throw, double-pole switch that would be used to eliminate the need for two switches in a circuit. It still uses the same number of wires and does the same job, but it takes up less space and at a lower cost for the OEM's.

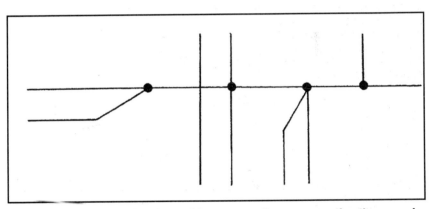

When lines go across other lines without any splice or connection they remain uninterrupted, but if they make a connection or a splice with another wire in their path they are marked with a dot. All but one of the lines in the second example makes a connection. Note how there are two cases where parallel lines make an angle and connect to another line. Generally these would be where a power line gets split or where grounds join to larger wires on the way to a common ground.

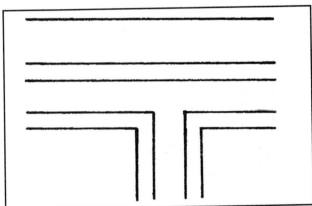

It is easy to see how straight lines can be drawn at right angles, but lines at obliques (less than 90 degree angles) are also part of the mix if that angled line delivers some form of useful information to the diagram reader. In most cases straight lines and lines at right angles are preferred. The diagram of the F.A.S.T./EFI system at the start of this chapter used only straight lines and right angles and it is very well drawn and informative.

transmission. The car we worked on is actually a 1972 and was heavily optioned as it came out of the factory. But by the time the present owner bought it most of the options were gone, removed for drag racing. The diagrams came from Jim Olson a publisher of wiring diagrams for many of the cars built in the 1950s, '60s and '70s.

Look at the diagram for the car's center section on page 46 (a larger view is on page 91). It shows all of the electrical components that are in the center of the car—underdash instruments and heater controls, all light controls and accessories. In other words, everything the driver will interface starting at the fuse panel up to where it leaves the cabin at the bulkhead connector or body connector. The owner of the vehicle rewired the car using OEM replacement wires from YearOne.

Now look at the diagram again to locate the fuse panel, bulkhead disconnect (bulkhead connector) steering column, radio, accessory feeds and finally

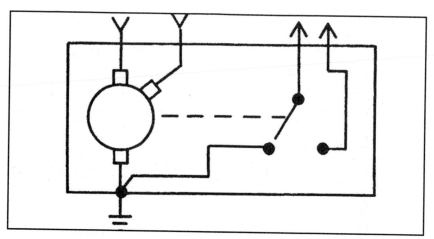

This is the schematic for an electric wiper motor and washer system. This shows that it is multispeed and intermittent. This is what you will find on most modern vehicles today. They offer so many benefits for the driver that the factories have quit making the simple two-speed type. There would also be another motor to show the washer water pump, where it is located and how the circuit gets activated and grounded.

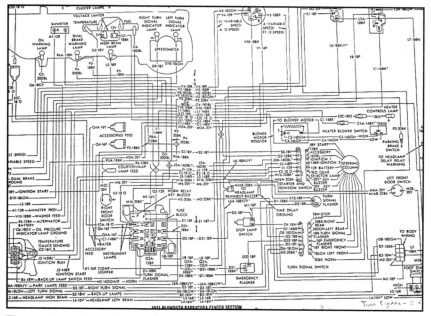

The center section diagram is very comprehensive, displaying all electrical components, wires, switches, relays located under the dash, and all lights, heater, washer/wiper controls and more. See what you can find and trace things like the radio and starter circuits back to their power source. For a larger view of this diagram, turn to page 91.

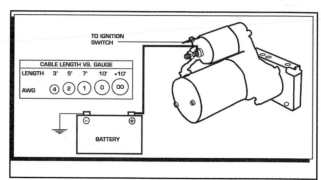

This wiring diagram is for MSD's starter. It illustrates what needs to be done for the installation and is easy to follow.

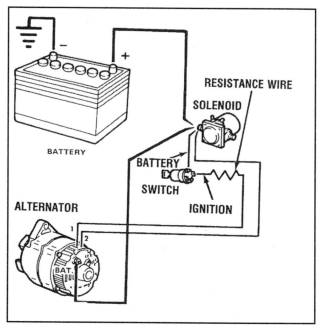

Simply diagramed circuits will be found in many teaching materials and included with most aftermarket electrical parts. They actually take part of a vehicle's circuitry out of context and use it as material for learning or a "how-to" installation. For example how simply could a parallel light circuit be drawn? You need nothing more than a battery, power wires, bulbs and a wire going to ground before the battery and after the bulbs. But, if we want to show a student a more complete diagram of a starter circuit, for example, we would use the wires needed but replace symbols for the components with drawings that look more realistic to help the student understand.

the light circuits. (You will find it on the right side of the diagram.)

Now take a look at that first diagram from F.A.S.T. again on page 43. By ignoring all the wires running through the middle, and reading everything around the outside, you will now find it much easier and less intimidating. Each sensor, injector, relay, fuse is clearly labeled and as you trace these wires around you will see which connector they go to as well as their color and pin location.

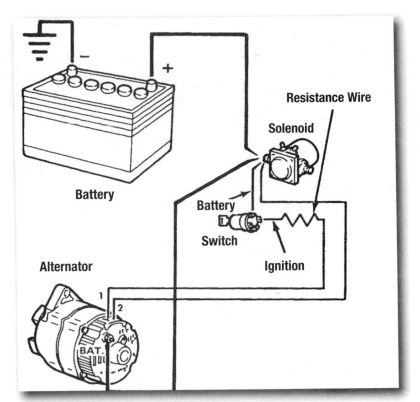

The charging system consists of the battery, the alternator, the wires that run between them and the belt system used to spin the alternator. The battery's main function is to store and deliver electricity to start the engine. Once running the electrical system, the alternator takes over. The starter and ignition switch only come into play during the starting cycle. After that the keyed switch is simply an electrical distribution buss.

Chapter 6

Primary Electrical Systems

This chapter consists of four sections providing information on the basic circuits that every car/vehicle must have to operate safely on public roads—electrical charging and storage, starting, lights and wipers. Although there are other systems you might consider critical, such as a heating and A/C or audio, these four are essential to make the vehicle run legally on the street.

THE CHARGING SYSTEM

The charging circuit's main function is to supply electricity to the vehicle while the engine is running and to store electricity to allow you to start the engine when it is off and to run accessories. In this section we will cover its components and how they work, and how to select replacement or performance components for your vehicle.

Battery

The lead acid battery is the heart of the charging circuit. It is an electrochemical device for producing and storing electricity. A vehicle battery has several important functions apart from the obvious one of storing electricity. It must operate the starting motor, ignition system, engine management system, electronic fuel injection system and other electrical devices for the engine during engine cranking and starting.

It supplies all the electrical power when the engine is not running. The battery will provide electricity when current demands are above the output capacity limit of the charging system. It also acts as a voltage stabilizer to smooth current flow through the electrical system, which is important since some components create spikes that might damage other components.

The type of battery used in automotive applications is a common lead-acid cell-type battery. This type of battery produces direct current (DC) electricity that flows in only one direction. When the battery is discharging (current flowing out of the battery), it changes chemical energy into electrical energy, thereby releasing stored energy. During charging (current flowing into the battery from the charging system), electrical energy is converted into chemical energy. The battery can thus store energy until the vehicle requires it.

The battery takes quite of bit of abuse in the engine compartment, so it must be built to withstand severe vibration, cold weather, engine heat, corrosive chemicals, high current discharge, and prolonged periods of inactivity without discharging too quickly. Most auto batteries are sealed with plastic covers, and are maintenance-free, meaning you don't have to fill them with water like older batteries. Inside that plastic case there are 6 separate cells to produce a nominal 12 volts. Each cell consists of battery plates, positive and negative, separators and electrolyte. The cells are connected together with straps in series and then

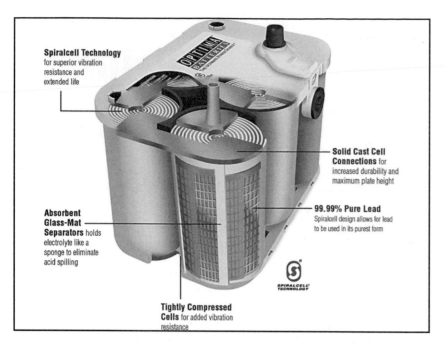

This Optima spiral cell technology is unique and patented, but looking at it makes it clear how each cell is separate and how the plates and separators are installed. This battery looks like 6 cans were joined at the top to make a 12 volt battery. That is essentially what a battery is. On the flat cell–style battery they also have 6 separate cells, but they are not wound up like the Optima. The Optima batteries are preferred by the racing community because they are tougher than most regular flat designs. Some of Optima's batteries can be mounted vertically or even upside-down because they use an acid gel, not a liquid.

Inside the durable plastic case of a modern lead-acid battery you will find positive and negative plates with separators keeping them from shorting out. If we had to endure what the battery does (heat, cold, vibration, bumps, etc.) we wouldn't last long. The modern battery is a wonder of technology and quality.

everything is neatly packaged and protected by the case. In Chapter 1 we discussed how important it is that they are connected in series because each cell produces a nominal 2 volts, so six cells equal basically 12 volts. But if you plan to run two batteries, you would not wire them in series, because that delivers 24 volts and can damage or destroy just about everything electrical in the vehicle. You would wire them in parallel so you would always have 12 volts, but the second battery would double the available amperage to meet high-load demand, such as when you crank that 2,000-watt stereo or hit the rack of driving lights overhead.

Battery Construction—Each cell compartment contains two kinds of chemically active lead plates, known as positive and negative plates. The battery plates are made of GRID (stiff mesh framework) coated with porous lead. These plates are insulated from each other by suitable separators and are submerged in a sulfuric acid solution (electrolyte). Charged negative plates contain spongy (porous) lead (Pb) which is gray in color. Charged positive plates contain lead peroxide (PbO_2) which has a chocolate-brown color. These substances are known as the active materials of the plates. Calcium or

antimony is normally added to the lead to increase battery performance and to decrease gassing (acid fumes formed during chemical reaction). Since the lead on the plates is porous like a sponge, the battery acid easily penetrates into the material, which increases the actual surface area by many times and therefore the efficiency of the battery.

Lead battery straps or connectors run along the upper portion of the case to connect the plates—positive on one side and negative on the other. They must never touch or the battery will short out. To prevent the plates from touching each other and causing a short circuit, sheets of insulating material (microporous rubber, fibrous glass, or plastic-impregnated material), called separators, are inserted between the plates. These separators are thin and porous so the electrolyte will flow easily between the plates. The side of the separator that is placed against the positive plate is grooved so the gas that forms during charging will rise to the surface more readily. These grooves also provide room for material that flakes from the plates to drop to the sediment space below.

The battery case is made of hard rubber or a high-quality plastic. The case must withstand extreme vibration, temperature change and the corrosive action of the electrolyte. The dividers in the case form individual containers for each cell. Stiff ridges or ribs are molded in the bottom of the case to form a support for the plates and a sediment recess for the flakes of active material that drop off the plates during the life of the battery. The sediment is thus

A battery can have both side and top post connections as shown here. Use either the ones on top or the side mounts and leave the plastic covers in place on the ones you don't use. This design also gives you another distribution buss for positive wire connections, but more importantly, for grounds.

kept clear of the plates so it will not cause a short circuit across them. The battery cover is made of the same material as the container and is bonded to and seals the container. The cover provides openings for the two battery posts and a cap for each cell. Battery caps either screw or snap into the openings in the battery cover. The battery caps (vent plugs) allow gas to escape and prevent the electrolyte from splashing outside the battery. They also serve as spark arresters (keep sparks or flames from igniting the gases inside the battery). The battery is filled through the vent plug openings. Maintenance-free or sealed batteries have a large cover that is not removed during its service life.

Caution—Hydrogen gas can collect at the top of a battery. If this gas is exposed to a flame or spark, it can explode! Use extra caution when working around batteries, wear protective clothing and eyewear.

Battery terminals provide a means of connecting the battery plates to the electrical system of the vehicle. Either two round posts or two side terminals are used. Battery terminals are round metal posts extending through the top of the battery cover. They serve as connections for battery cable ends. The positive post may be larger than the negative post. It may be marked with red paint and a positive (+) symbol. Some negative posts are smaller, may be marked with black or green paint, and have a negative (–) symbol on or near it. Side terminals are electrical connections located on the side of the battery. They have internal threads that accept a special bolt on the battery cable end. Side

terminal polarity is identified by positive and negative symbols marked on the case.

The electrolyte solution in a fully charged battery is a solution of concentrated sulfuric acid in water. This solution is about 60 percent water and about 40 percent sulfuric acid. Hopefully you will get nowhere near it. Respect the dangers acid can represent and do things the right way. The electrolyte in the lead-acid storage battery should have a specific gravity of about 1.28, which means that it is 1.28 times as heavy as water. The amount of sulfuric acid in the electrolyte changes with the amount of electrical charge as does the specific gravity of the electrolyte. A fully charged battery will have a specific gravity of 1.28 at 80°F. The figure will go higher with a temperature decrease and lower with a temperature increase. As a storage battery discharges, the sulfuric acid is depleted and the electrolyte is gradually converted into water. The resulting change in specific gravity downwards provides a guide in determining the state of discharge of the lead-acid cell.

Measuring Specific Gravity—The specific gravity of an electrolyte is actually the measure of its density. The electrolyte becomes less dense as its temperature rises. A hydrometer is used to test the specific gravity where there is access. It will be marked to read specific gravity at 80°F only. Under normal conditions, the temperature of your electrolyte will not vary much from this mark and so normally the temperature is not important. However, large changes in temperature require a correction in your readings. For every 10-degree change in temperature above 80°F, you must add 0.004 to your specific gravity reading. For every 10-degree change in temperature below 80°F, you must subtract 0.004 from your specific gravity reading.

Suppose you have just taken the gravity reading of a cell. The hydrometer reads 1.280. A thermometer in the cell indicates an electrolyte temperature of 60°F. That is a normal difference of 20 degrees from the normal of 80°F. To get the true gravity reading, you must subtract 0.008 from 1.280. Thus the specific gravity of the cell is actually 1.272. A hydrometer conversion chart is usually found on the hydrometer itself. From it, you can obtain the specific gravity correction for temperature changes above or below 80°F.

The capacity of a battery is measured in ampere-hours. The ampere-hour capacity is equal to the product of the current in amperes and the time in hours during which the battery is supplying current. The capacity of a new cell, and therefore the battery, depends upon many factors. The most important of which are the surface area of the plates in contact with the electrolyte (acid), the quantity and specific gravity of the electrolyte and the type of separators.

As the battery ages, its general condition becomes a factor. The degree of sulfating, whether the plates are buckled or the separators warped, how much sediment there is in the bottom, are all factors. Some of these will eventually lead to the battery failing completely. In any case, as one or more cells fail, the output voltage will fall. Under normal conditions, a hydrometer reading below 1.240 specific gravity at 80°F is a warning signal that the battery should be removed and charged. Except in extremely warm climates, never allow the specific gravity to drop below 1.225. In general all of this isn't really important to you—all you want is to know if the battery is good or bad. They sell simple hydrometers that will give you a go or no-go reading for very little money. Keep one in your toolbox.

Charging a Dead Battery—When you have identified a rundown battery, you should recharge it immediately. Use a good make of battery charger. For safety it is better to disconnect the battery and remove it from the vehicle to protect key components. However some chargers are safe to use on the vehicle. The charger's instructions should be able to guide you on whether you can leave the battery connected and/or installed. If you are going to do it in the car you should at least remove the battery cables.

You should always disconnect and remove a battery from the vehicle during charging. It is a little more work, but having it isolated gives you more control and protects the vehicle's circuits. Connect the battery to the charger before you plug in the charger and turn the charging current on. The charger does the rest. Before you try to remove the cables, unplug the charger. After a normal charging period, or when the charger tells you the battery is fully charged, turn the charging current off, unplug it and remove the cables from the battery in that order. Certain precautions however are necessary both before and during the charging period.

1. Clean and inspect the battery thoroughly before placing it on the charger. Use a solution of baking soda and water for cleaning; and inspect for cracks or breaks in the container.

2. If the battery is not a sealed type, check the electrolyte level before charging begins and during charging. Add *distilled* water if the level of electrolyte is below the top of the plate.

3. Before plugging in the charger, connect the battery to the charger. Be sure the battery terminals

Although this is just a small motorcycle battery being charged, it is exactly the same as charging a big vehicle battery. Isolate the battery, clean it, connect the cables and then plug it in. If the battery isn't good it will not charge. This new motorcycle battery replaced the one that was in it and wouldn't take a charge. The charger told me (using the screen) that it was sulfated and unusable, therefore the new battery.

are connected properly; connect positive post to positive (+) terminal and the negative post to negative (-) terminal. The positive terminals of both battery and charger are marked; those unmarked are negative. The positive post of the battery is, in most cases, slightly larger than the negative post. Ensure all connections are tight.

4. See that the vent holes are clear and open. Do not remove battery caps during charging. Acid can spray onto the top of the battery and the caps keep dirt out of the cells.

5. Keep the charging room well ventilated. Do not smoke or light anything with a flame near a battery. Batteries on charge release hydrogen gas. A small spark may cause a big explosion.

6. Take frequent hydrometer readings of each cell and record them. You can expect the specific gravity to rise during the charge. If it does not rise, the battery is no longer accepting a charge and is at the end of its life. Replace the battery and dispose of it as per local hazardous material disposal instructions.

7. Keep a close watch for excessive gassing, especially at the very beginning of the charge when using the constant voltage method. Reduce the charging current if excessive gassing occurs. Some gassing is normal and aids in remixing the electrolyte.

8. Do not remove a battery until it has been completely charged and the charger has been unplugged.

Caution—When you remove the battery always clean the entire mounting and hold-down assembly using a solution of baking soda and water that will neutralize any residual acid before reinstalling the battery. Do not allow the solution to enter the cells while you are cleaning away the corrosion, as it will then neutralilze the acid inside and destroy the battery. New batteries are usually sold full of electrolyte and fully charged. In this case, all that is necessary is to install the batteries properly. If your battery is not maintenance-free, you must spend some time on periodic maintenance or its service life will be drastically reduced. Complete battery maintenance requires you to check the battery visually for any cracks and for terminal corrosion problems. If you can, check the electrolyte and top it up with distilled water when necessary. A visual inspection of the battery should be done regularly.

Inspection—Look for signs of corrosion on or around the battery, signs of leakage, a cracked case or top, missing caps, and loose or missing hold-down clamps. On vent cap batteries the electrolyte level can be checked by removing the caps. Some batteries have a fill ring, which indicates the electrolyte level. The electrolyte should be even with the fill ring. If there is no fill ring, the electrolyte should be high enough to cover the tops of the plates. Some batteries have an electrolyte-level indicator (Delco Eye). This gives a color-coded visual indication of the electrolyte level, with black indicating that the level is okay and white meaning a low level.

If the electrolyte level in the battery is low, fill the cells to the correct level with distilled water (purified water). Distilled water should be used because it does not contain the impurities found in tap water. Tap water contains many chemicals that reduce battery life. The chemicals contaminate the electrolyte and collect in the bottom of the battery case. If enough contaminates collect in the bottom of the case, the cell plates short out, ruining the battery. If water must be added at frequent intervals, the charging system may be overcharging the battery. A faulty charging system can force excessive current into the battery. Battery gassing can then remove water from the battery.

Maintenance-free batteries do not need periodic electrolyte service under normal conditions. They are designed to operate for long periods without loss of electrolyte. But they must be clean and free of corrosion: use the same baking soda/water solution described above to neutralize the acid. To clean the terminals, remove the cables and inspect the terminal posts to see if they are deformed or broken. Clean the terminal posts and the inside

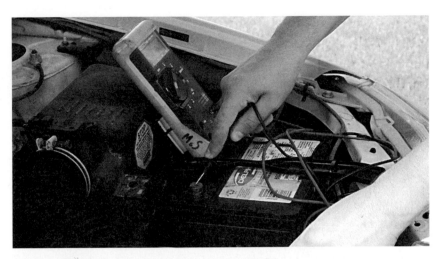

Using a multimeter to test voltage drop is a quick and sure method of testing it. (The hydrometer is generally faster and you will not use it often and probably will not buy one either.) Voltage drop is what you need to look for.

surfaces of the cable clamps with a wire brush cleaning tool before replacing them on the terminal posts. After reinstalling the cables, coat the terminals with petroleum or white grease, which will keep acid fumes off the connections and keep them from corroding again. Tighten the terminals just enough to secure the connection. Over-tightening will strip the cable bolt threads.

Caution—Do not use a scraper or knife to clean battery terminals. You can ruin the terminal connection. Instead use a wire brush or a battery-terminal brush.

Checking Battery Current—Electrical problems drawing current out of the battery with the ignition off can create a defective battery and can be found by using a hydrometer to check each cell. If the specific gravity in any cell varies excessively from other cells (25 to 50 points), the battery is bad. Cells with low readings may be shorted. When all of the cells have equal specific gravity, even if they are low, the battery can usually be recharged.

On some maintenance-free batteries a charge-indicator eye shows the battery charge. Others have nothing to judge condition with. The charge indicator changes color with levels of battery charge. For example, the indicator may be green with the battery fully charged. It may turn black when discharged or yellow when the battery needs to be replaced. If there is no charge-indicator eye or when in doubt of its reliability, a voltmeter and ammeter or a load tester can also be used to determine battery condition quickly.

Alternators

In modern vehicles built from the '70s on, the charging system will be an alternator. Alternators

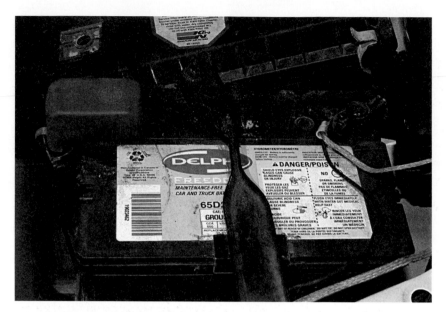

Maintenance-free batteries have been around for many years. All you need to do is test them when there is a charging system problem by doing a voltage drop and replace them if they are bad.

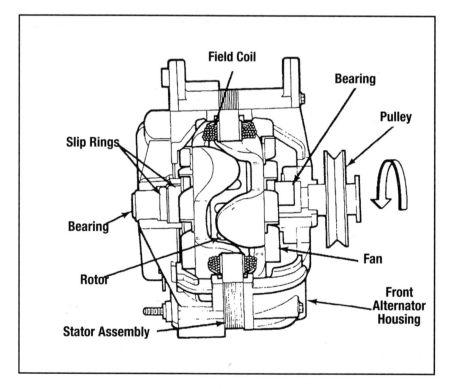

Field Coil

Bearing

Pulley

Slip Rings

Bearing

Rotor

Fan

Front Alternator Housing

Stator Assembly

This is a typical AC alternator. Knowledge of the construction of an alternator is required before you can understand the proper operation, testing procedures, and repair procedures applicable to an alternator.

deliver almost maximum power from tick-over speed to high rpm, provide a cleaner supply and are more powerful than generators. It is smaller, lighter, and more dependable than the DC generator. The alternator also produces more output at idle which makes it ideal for late-model vehicles. The alternator has a spinning magnetic field. The output windings (stator) are stationary. As the magnetic field rotates, it induces current in the output windings.

Rotor Assembly—The rotor consists of field windings (wire wound into a coil placed over an iron core) mounted on the rotor shaft. Two claw-shaped pole pieces surround the field windings to increase the magnetic field. The fingers on one of the claw-shaped pole pieces produce south (S) poles and the other produces north (N) poles. As the rotor rotates inside the alternator, alternating N-S-N-S polarity and AC current is produced.

An external source of electricity is required to excite the magnetic field of the alternator. Slip rings are mounted on the rotor shaft to provide current to the rotor windings. Each end of the field coil connects to the slip rings.

Stator Assembly—The stator produces the electrical output of the alternator. The stator, which is part of the alternator frame when assembled, consists of three groups of windings or coils that produce three separate AC currents. This is known as three-phase output. One end of the windings is connected to the stator assembly and the other is connected to a rectifier assembly. The windings are wrapped around a soft laminated iron core that concentrates and strengthen the magnetic field around the stator windings.

Rectifier Assembly—also known as the diode triad assembly, consists of six diodes to convert stator AC output into DC current. Diodes allow current to flow in only one direction. So the bridge is set up to allow the passage of positive flow of electricity but when the flow is in the negative part of the cycle, it grounds it. A diode allows positive current to flow from the winding as it is allowed to pass through. As the current reverses direction, it flows to ground through a grounded diode. The insulated and grounded diodes prevent the reversal of current from the rest of the charging system.

The number of pulses created by motion between the windings of the stator and rotor keeps a fairly even flow of current is supplied to the battery terminal of the alternator. The rectifier diodes are mounted in a heat sink (metal mount for removing excess heat from electronic parts). The whole assembly is usually called a diode bridge. Three positive diodes are mounted in an insulated frame.

Three negative diodes are mounted into a grounded frame. When an alternator is producing current, the insulated diodes pass only out flowing current to the battery. The diodes provide a block, preventing reverse current flow from the alternator.

Alternator Operation—Since the engine speed varies in a vehicle, the frequency also varies with the change of speed. Likewise, increasing the number of pairs of magnetic north and south poles will increase the frequency by the number pair of poles. A four-pole generator can generate twice the frequency per revolution of a two-pole rotor.

Alternator Output Control—Alternator speed and load determine whether the regulator increases or decreases charging output. If the load is high or rotor speed is low (engine at idle), the regulator senses a drop in system voltage. The regulator then increases the rotor's magnetic field current until a preset output voltage is obtained. If the load drops or rotor speed increases, the opposite occurs.

Alternator Maintenance—In most cases you will never have to worry about the alternator. Most will run well over 150,000 miles with the same alternator. But, you must make sure it is not covered with dirt or mud. The main thing is to make sure that the drive belt is in good shape and the tension on the belt is correct.

Alternator maintenance is minimized by the use of pre-lubricated bearings and longer-lasting brushes. If a problem exists in the charging circuit, check for a complete field circuit by placing a large screwdriver on the alternator rear-bearing surface. If the field circuit is complete, there will be a strong magnetic pull on the blade of the screwdriver, which indicates that the field is energized. If there is no field current, the alternator will not charge because it is not excited by battery voltage. Should you suspect troubles within the charging system after checking the wiring connections and the battery, connect a voltmeter across the battery terminals. If the voltage reading, with the engine speed increased, is within the manufacturer's recommended specification (usually between 13.0 volts to 14.4 volts) the charging system is functioning properly. Should alternator tests fail, the alternator should be removed for replacement. Do not forget, you must always disconnect the cables from the battery first. Alternator testing and service call for special precautions since the alternator output terminal is connected to the battery at all times. Use care to avoid reversing polarity when performing battery service of any kind. A surge of current in the opposite direction could burn the alternator diodes. Do not purposely or accidentally short or ground the system when

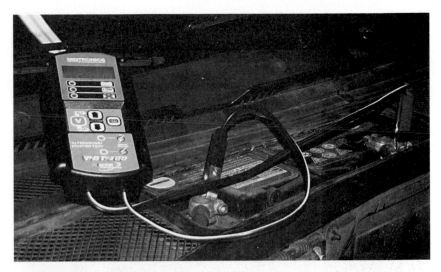

Testing is always called for before changing components. I've seen vehicle owners remove their battery and alternator and take them to an autoparts store. The store will test them and only sell them a new part if one of theirs is bad. Driving the entire vehicle there if it will run is easier, but not always possible. This tester analyzes the entire charging system.

disconnecting wires or connecting test leads to terminals of the alternator or regulator. For example, grounding of the field terminal at either alternator or regulator will damage the regulator. Grounding of the alternator output terminal will damage the alternator and possibly other portions of the charging system. Never operate an alternator on an open circuit; without electrical load in the circuit because alternators are capable of building high voltage (40 to over 110 volts), which may damage diodes and endanger anyone who touches the alternator output terminal.

Alternator Drive Belt System—The alternator will be driven off the engine with either a V-belt or a serpentine belt system. If it is a V-belt, keep an eye on the engine drive pulley, alternator pulley and the belt. Make sure the pulleys are aligned properly, and that the belt is set at the correct tension and is in good condition. Check it periodically for cracks or fraying. A new belt is cheap insurance against being stuck on the highway.

With a serpentine belt that drives many components (water pump, power steering, A/C), make sure everything is in good shape (especially the water pump which seems to go bad at about 75,000 miles on modern cars); I change belts about every 35,000 miles just to make sure they are perfect. Belts need to be replaced when they show signs of cracking, excessive wear on the drive side of the belt or signs that they have been slipping. On serpentine belt systems the tension control should be replaced or adjusted at about 75,000 miles.

The charging system of any automotive

If the alternator is driven with a V-belt as shown here, make sure it is in good shape.

Heavy-duty, high-amperage alternators tend to wear out faster, usually at about 75% of the stock unit. When you need the extra power, spend the extra money for a good one.

When the alternator is on top of the engine it is far easier to get to and test. You can always tell if it is an alternator or a different accessory because the alternator always has the open areas for cooling that show windings.

application is simple and easy to diagnosis and repair. Just remember that the first thing to do is to look at the grounds. After that, if they are good, you can start looking at drive belts and other things. Remember an alternator can only work when it is turning, so do not forget to check the drive belt first. After a good visual inspection you should test the battery, alternator and the wires between them. Replace what is bad and you will be good to go.

THE STARTING SYSTEM

Before you can do any testing on the starter system, you must make sure the battery and wires are in good working order. Remember, the battery's

only job is to store and supply electrical energy to start the engine so the alternator can take over the chores of electrical supply. Review the section on the charging system in this chapter, then test the battery and wires leading to the interior of the car and the key as well as the wires running from the key to the starter solenoid. There should always be a big positive cable on the starter solenoid. If it is there, and it has 12 volts current going through it, you can be sure it is probably fine, if the battery is also fine.

Troubleshooting the Starter

One of the things I do when the engine will not turn over is to listen for the solenoid to engage and to hold on to the engine ground. If the solenoid engages and the engine ground wire gets hot you have a weak engine ground that will stop the starter from working properly. It will also blow out fuses. Again, grounds are the major culprit in vehicle electrical system problems. Make sure the engine ground isn't loose, broken, corroded or undersize. If this was the first start-up on a new engine, and the ground wire was getting hot, I would look to see if the ground was fastened to the motor mounts because they are rubber insulated and will not conduct current to ground.

Although your problem could be the alternator, battery or starter motor (or in the wires connecting them), getting a jump start should have told you which system to look at first. If you hadn't been able to get it to start with a jump start, the starter

Regardless of the type of starter motor in your vehicle, it contains the only mechanical components that are needed to start the car. The first is the starter solenoid and the second is the gear the starter motor shoots out to engage the starter ring on the flywheel that will turn the engine over. It is really a simple system to diagnose and repair. That is, unless the starter ring on the flywheel is broken, loose or badly damaged. Then you'll need to change the flywheel or flexplate, which can be easy or difficult depending on the vehicle.

	3'	5'	7'	10'	15'	20'	25'
5 Amps	18	18	18	18	18	18	18
6	18	18	18	18	18	18	18
7	18	18	18	18	18	18	18
8	18	18	18	18	18	16	16
10	18	18	18	18	16	16	16
11	18	18	18	18	16	16	14
12	18	18	18	18	16	16	14
15	18	18	18	18	14	14	12
18	18	18	16	16	14	14	12
20	18	18	16	16	14	12	10
22	18	18	16	16	12	12	10
24	18	18	16	16	12	12	10
30	18	16	16	14	10	10	10
40	18	16	14	12	10	10	8
50	16	14	12	12	10	10	8
100	12	12	10	10	6	6	4
150	10	10	8	8	4	4	2
200	10	8	8	6	4	4	

I have placed this chart in this section because starter circuits are where a downsized wire size will be fried faster than anywhere else. At left are the amps, at top is the distance and numbers in the shaded areas indicate the wire gauge needed.

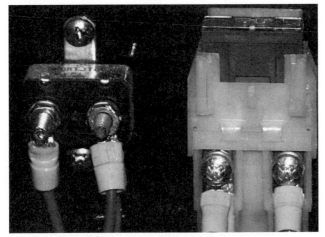

If the starter system has a bad ground or a short to ground, the master fuse is probably blown. If it is, find the ground or short problem before you replace the fuse and try to start it again. Look for a bad ground first and then test the main wires from the ignition switch to the solenoid to the starter for problems. It might be something as simple as a battery wire that has chafed. Look for the simple first, as always.

system, not the charging system, would have been the place to look. Let's go forward as if the engine wouldn't start with a jump from AAA. If you know the battery is perfect, the first test is to see if that electricity is reaching the big positive cable at the starter. If your multimeter says the starter is getting +12 volts, your next step is to see if the key is sending electricity to the solenoid. Have someone turn the key as you use the multimeter to check for power at the small starter solenoid terminal (usually the one on the left). If you get a 12-volt reading at the terminal you can forget the switch as far as the starter is concerned. Have your helper turn the key off and make sure the power disappears. If it doesn't, test another wire.

Run a ground wire from the body of the solenoid to a good ground (if your solenoid is self-grounding as most are) and have someone turn the key to start again. If the engine turns over, you have found the problem—the solenoid needs to be grounded. A bad ground of the starter motor, either on the motor or the bellhousing depending upon which it is mounted to, will also keep it from starting. This is a simple problem, so check for it early. If it is a Ford, the solenoid needs to be checked where it is mounted on the inner fender or firewall. It is probably a self-grounding unit also. If it happens to have a safety switch built in, make sure it is closing.

The size of the engine and the compression ratio, type of induction, and accessories being turned by the engine can all gang up on the starting system and make it hard to turn over. If your engine has a supercharger on it you might be surprised to find that it will take about 45 hp to turn it at low speeds, and to get it started turning along with a

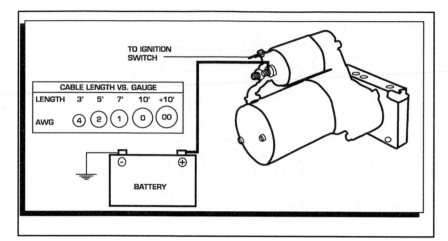

Take a look at the starter circuit in diagram form, because seeing a simple visual representation of the circuit will allow you to understand it better. There is the battery, starter and ignition switch. Since most solenoids are mounted on the starter itself (except for Fords), it makes it easy to find everything—nothing complicated. All the key ignition switch does is send current to the starter solenoid (small connector) to set the relay into action. If the starter system worked yesterday and didn't last night, nothing too complicated has happened. You will probably find the problem with either a visual inspection or a battery test. Courtesy MSD Ignition.

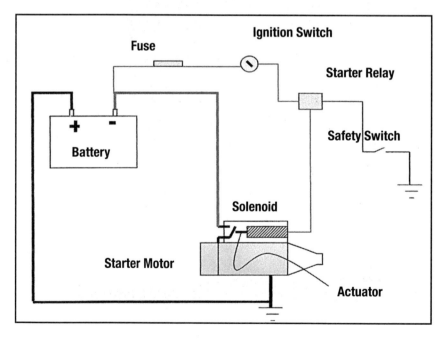

This starter system diagram is a little more inclusive and shows you where a fuse is located in a schematic form. Once again that fuse is probably between 75 and 100 amps and should be checked early in the diagnosis process. This diagramed system is like the Chrysler system that has a separate solenoid mounted on the fenderwell or firewall, but has the actual plunger/actuator on top of the starter. Note that the starter is the ground for the plunger/actuator, yet it also has a safety switch to ground the solenoid.

high-compression engine is a real load. Additionally, the power steering can load another 15 to 20 hp on the engine and if you leave the A/C on you are talking about another 25 hp. That is 90 hp required to turn that engine over, not counting inherent drag by the pistons, rings, cam(s), etc. Since the engine isn't running, the starter has to generate that 90 hp if you want the engine to start. If the vehicle came with all these accessories OEM, the stock battery was probably sized to do the job. But, after a few years the entire system starts getting corroded and things may simply require so much power to turn over that the battery gets drained and/or will not turn over at all. Consider all of these things as part of your diagnostic strategy.

If you purchased your vehicle used, and it was modified with aftermarket components, you may find all sorts of surprises during the first few weeks of ownership. Things might fall off, it may overheat quickly (just past the distance you used on your test drive) and a lot of things that were on the car from the factory might have been removed (just as bad as having too many things added). I have seen so many horror stories, and lived through a few of them in my earlier years, that these things never surprise me anymore. I've seen people install a 500 hp new engine without consideration about how they plan to stop the thing with stock brakes. I've seen people install 5,000 watts of audio/video equipment and the first time out show off to their friends with the engine off by trying to break the windows out of Macy's on Fifth Avenue. Much to their surprise the thing will not start after their 20-minute demo; they have a dead battery and nothing to get it started. (Sure hope they have a road service policy.) The same thing can happen on simple things like engines, starters, and other things people modify without a full understanding of the electrical systems in a vehicle.

Do not be surprised if you have to go behind the dash to look at what changes were made by the last owner. Anytime someone goes behind the dash to locate a hot lead keyed for some accessory, the chances are high they will do something wrong. This is especially true as it pertains to the starter system. Since the key also sends electrical energy to the solenoid, it is a likely place for the addition of aftermarket accessories to find keyed current. The problem with this happening is that electricity always selects the path of least resistance. If the accessory operates all the time the key is on, it might pull a lot of power from the circuit and that power might be needed for the starting system.

Any time there is an electrical problem on a used vehicle that has been modified, look at the changes

This Toyota Corolla is a 4-cylinder engine (top) created to deliver outstanding fuel economy and was designed from scratch to have very little internal and parasitic drag (resistance to turning). It can get by with a 50-amp alternator and a small battery to operate well. The big-block Chevy 454 (bottom) has high-compression, tight rings, strong valve springs, flat-tappet hydraulic lifters and inherently has a lot of parasitic drag. These differences can make a dramatic difference in what it takes to turn over the engine. This engine needs an 85–110-amp alternator, plus a gear-reduction starter to turn this monster over enough for it to start.

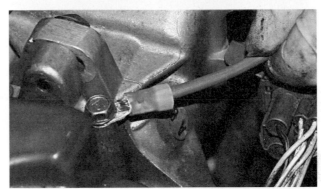

The engine ground wire on this Acura 4-cylinder engine was attached to the valve cover nut (top) and stud with a flat washer insulated with rubber to help stop oil leaks. It was a good enough ground to start the engine, but the engine wouldn't rev above 5,000 rpm, which isn't where these hot little engines were designed to operate. The ground wire also got hot when starting the engine after turning laps at a local race track. By changing to a #10 gauge wire and grounding it to the head at the distributor mounting area (bottom), the engine would pull beyond the 7,500 rpm range and deliver its full power potential. It also started easier and the wire itself never got hot again. There are many other places where starting systems can go bad, and not all of them are quite so simple to find after the vehicle is assembled.

made by the owner before looking at other things. The battery box installation above is typical of the ways aftermarket parts can be installed wrong. Just take the time to look things over and make any corrections needed. Then when problems arise, you will know that things were right before the problem. If the positive lead of the truck had been installed with the aluminum box touching, it would take very little time before the metal could chafe it and short out the system. It would take a lot of work to locate the problem unless you knew what to look for. The bottom line on the starter system is that it is simple and easy to troubleshoot. Very few mechanical things are involved and the wiring is simple. Be patient and you will diagnose and fix it quickly.

LIGHTS

Every time you reach over and turn on the lights of your vehicle you expect the headlights, taillights, side marker lights, instruments lights and the license plate lights to turn on. You take it for granted that these lights will turn on and do their job of providing light for you to drive by and for the other lights to help another vehicle to identify you as a car/vehicle. If these things do not happen, you will be in a dangerous position, either driving

Headlight systems, as well as brake and signal circuits, must be functional and reliable for any street-legal vehicle. Make sure you spend the extra time to install and maintain these circuits.

This dash light switch allows the driver to turn on the running lights with or without the headlights. Most states require you to have all lights on anytime any of them are turned on. You can also see that the instrument light dimmer rheostat is just to the right of the switch. In this vehicle the dimmer controls dash/instrument lights and also allows the driver to turn on the interior lights without having to open a door.

On top is an old foot-operated high beam dimmer switch from a 1972 Barracuda. You can see a lot of dirt and moisture built up near the contacts. It was close to shorting out, which would have burned up the wiring—a common problem with these switches. At bottom is a column-controlled dimmer switch. There is little that can go wrong with this switch—dirt and moisture aren't a problem.

blind and/or invisible.

Most vehicles have either a dash switch that turns on the lights or a switch on the steering column that does the same thing. There will also be a switch that allows you to turn on your bright lights—when no other vehicles are coming towards you. This switch is combined with the light switch on the steering column in many cases. The steering column may also be where the high beam switch is located even if the headlight switch is in the dash.

On older vehicles the high beam switch was on the floor where the driver could use a foot to turn on/off the high beams. This was a viable position for it, but it could also create a problem if the driver entered the vehicle with a lot of dirt, moisture or even snow and ice on his feet. After a time this could short out the switch and that usually led to burned wires as well as a fuse blown. When the OEM's moved the switch out of harm's way a lot of technical people said "It's about time."

Most of the lights in the vehicle have a single filament that produces light when electricity is passed through it. These would include all the interior lights, instruments and marker lights outside. But outside the front and rear park/turn signal lights have two filaments as do the headlights. By installing two filaments, a single bulb and holder can be used instead of two—this

represents a major cost reduction for the OEM's. Today, the OEM's are combining circuits every place they can, especially when it is in reference to accessories. Brake lights are actually on another circuit in most cases, turned on by the brake light switch operating off the brake pedal.

All light circuits in the vehicle are parallel so that a full 12 volts reaches each bulb. In the two drawings on page 59 you can see why the parallel circuit installation is critical. If it was a series circuit the bulbs would only get 6 volts, not enough for them to do their job. Never use a series circuit in any automotive system. Some aircraft, boats and commercial vehicles will use a 24-volt system by combining two 12-volt batteries in series, but the rest of the circuits are either modified for 24 volts, or there is a step-down transformer that reduces the voltage back to 12 volts. (Or the 12-volt power for the system can be taken from just one battery and they would not need a step-down transformer at all.)

As mentioned earlier, the brake light circuit is separate so there are no brake lightbulbs shown in chassis electrical/lighting diagrams. The turn

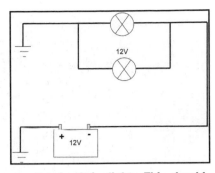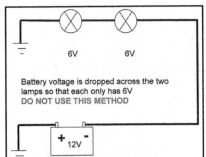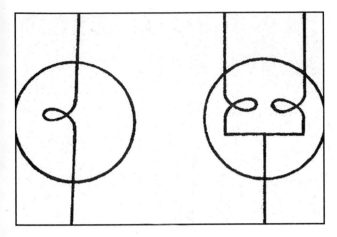

This is how single and double filament bulbs are diagramed in a typical electrical schematic.

Here are diagrams of both a series and a parallel circuit for lights. This should make it clear why parallel circuits are used on vehicles. There are no series circuits on an automotive vehicle and should never be used.

indicator lights on the dash are single filament and should be included because they are what tells the driver they are either working or not. When they quit working, the cause is usually a burned-out bulb. Flasher units will burn out, but if they do, both sides of the signal lights will not work, not just one, which indicates a bulb problem. Many owners add driving lights to their vehicles, especially trucks. All states require that they turn off when the high beams are turned on. There are also off-road lights available that are not for use on public roads. I have a set of these on my pickup because I go off-roading on a regular basis. Since these are not street-legal, they must be wired separately from the main headlight system.

The Chevy S-10 from 1997 uses rectangular headlights that are halogen while a 2005 Equinox uses much more modern and higher-performing headlights. These small lightbulbs use high-technology reflectors to focus their output to provide outstanding visibility for the driver. Hitting the high beams really lights up the world, but they are also less prone to blind oncoming drivers due to their improved focusing.

Other light circuits on a vehicle include running/side marker lights, turn signal lights and back-up lights (they are a separate light circuit from the headlights but integrated into the taillight housings). Between 1997 and 2005 (just 8 years) significant improvements were developed. Taillights on later model cars are usually LED bulbs, especially good for brake lights as they come on instantly compared to regular light systems. LEDs are usually bundled in a group, but if one burns out, the others continue to work.

WINDSHIELD WIPERS
When it comes to windshield wiper electrical

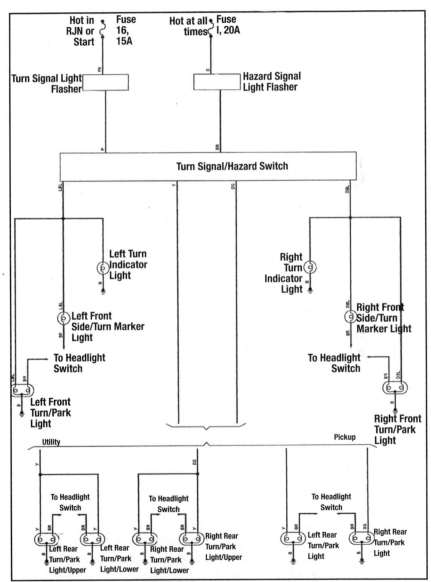

Looking at a diagram of the lighting circuit on a 1997 S-10 you will see which bulbs are single filaments and which are double. Note that there are lines with a point in the middle between the top part of the diagram and the bottom where the taillight bulbs are shown. This indicates the separation between the interior of the vehicle and the exterior. In most cases this will actually be a bulkhead connector doing the separator job. Courtesy Autozone.

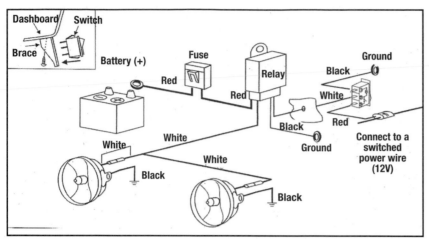

This diagram shows just how simple off-road lights are to install. The hardest parts of the job are locating where you want them installed and the mounting. Once they are installed it is time to run the wires. Since I know just how big a problem grounds can be, I ran the grounds for the lights all the way from the bumper to the battery ground and the power wire directly to the battery also. That made it a single ground system and easier to troubleshoot if they quit working. See page 84 for a step-by-step installation. Courtesy KC Hilites.

Don't forget about the license plate lights, a requirement in all fifty states. These illuminate the plate so it can be read while not generating enough light to bother driver's eyes.

circuits, there is relatively little to worry about. But, if you are reading this chapter it is probably because you want to fix your own wiper system. To get it fixed you need to define the problem and figure out the possible causes. The only wires are inside the vehicle under the dash, in most cases on early vehicles, and in the steering column wiring on new vehicles. Sure, it seems complicated, but it isn't if you take it one wire at a time. Using a test light you will be able to make sure each element of the circuit is getting power when that function is selected by the "wiper on" switch.

Typical Problems

The following are some of the problems followed by some suggested causes and solutions:

• *Will not turn on at all.* Check fuses, grounds, electricity from the fuse panel, wires to and from the switch, motor, junk in mechanical components stopping motor functions.

• *Wipers operate but windshield washers do not.* Check grounds, power to washer pump motor, hoses, water spray nozzles clogged.

• *Intermittent function doesn't work.* Check grounds, "power-in" wire, wiper motor assembly, wiper switch.

Other functions do not work, but low speed does. Check power, ground(s), wiper switch, wiper motor assembly, switch.

This modern wiper and washer system is mounted to the dash outside the cowl area. It has been developed to replace the vacuum motor for an early Ford pickup. It is wired to deliver all the same functions of the wiper system in any modern vehicle. There are four wires going to the switch and one of them is the power wire.

Early electrical vehicle wipers, once they eliminated the vacuum motors, have been self-contained inside the wiper motor assembly. Why? Because it makes installation easier on an assembly line to do it this way. Also, since the desired function will be performed by the wiper and washer assembly, having things close together makes all the sense in the world.

If you look at the wiper circuit on any interior wiring diagram you will find several switch positions and wires from the switch to the bulkhead connector. For example, on the 1972 Barracuda you can find six wires, plus one ground, coming off of the switch in the dash. Four of these go to the wiper motor assembly, and two go to the fuse panel. The

This shot of a modern vehicle's wiper switch shows just how far we have come in the last 30 years. Besides just turning the wipers on, this system includes an "intermittent-on" position that allows the driver to set the frequency they wipe the windshield (useful when the rain is light) plus "Lo," "Medium" and "High" settings. The same switch operates the windshield washer system, for the front and rear, plus turns on the rear wiper to clear the back glass on this SUV.

On most early vehicles the wiper system was powered by engine vacuum. These systems stuck around until the mid 1950s. The problem with vacuum wipers was that when you stepped on the gas they would quit working until the engine started making extra vacuum again. The vacuum wiper motor is right above the distributor mounted to the firewall.

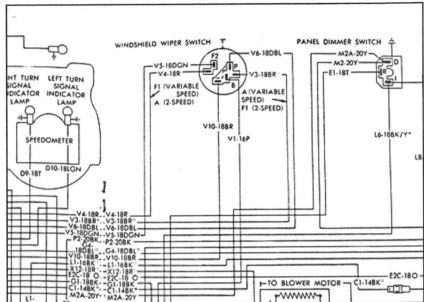

Looking at this interior wiring diagram shows that the only wiper system component inside the vehicle is the wiper switch. It has callouts showing this is a two-speed wiper motor and the windshield washer is included also. Six wires in the circuit isn't a big problem. You need to see if power is going through the switch in every position. That can be done at the switch connector.

Before you go too far with your troubleshooting, you should check the cowl where the wipers reside to make sure that leaves, twigs or other debris have not jammed up the motor or wiper mechanism. This is a common problem. If the mechanical part of the wiper system locks up it will add so much amperage draw it will burn out fuses. So make sure everything is clean and clear before you start the circuit testing process.

Remember, grounds are the most common problem in automotive electrics. Therefore, test the grounds at the switch inside the vehicle (under the dash) and at the motor before getting too involved with other tests.

Turn on the key to the run position and test to see if the switch has voltage-in from the fuse panel, verify that the wires at the switch connector have +12 volts. If they do not, the system cannot function. Look for problems at the fuse panel or the wires between the fuse panel and the switch. There should be +12 volts at the voltage-in side wires of the wiper switch, (you can test at the connector easily). If the voltage is available where it should be on the switch "voltage-in" side, the next step is to test it at the "voltage-out" side with the power applied to the switch.

The next step is to go to the "voltage-out" side (at the motor since it is easy to work on there). Disconnect the electrical connector and turn the

wires going to the wiper motor assembly end up at the bulkhead connector and those for the switch start at the vehicle fuse panel and end up powering the switch. The systems are just that simple. The wires going to the motor are for different speeds of the motor as well as activating the washer system.

From a diagnostic standpoint, the system is easy to test, using little more than a voltage test light and/or multimeter. But before starting the testing process, do a visual inspection of all connectors, the wires in general if visible, the fuses, bulkhead connections and the motor itself. (Look at the wiring as well as the mechanical parts of the motor and linkage assembly.)

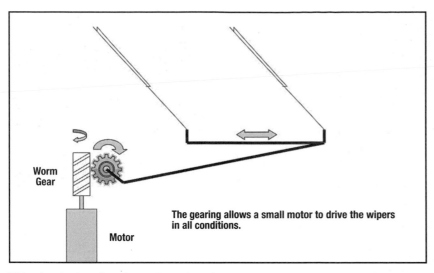

The gearing allows a small motor to drive the wipers in all conditions.

This simple drawing shows how the wiper motor turns a gear that drives the two wiper arms. It is a very practical way to build an electrical wiper system.

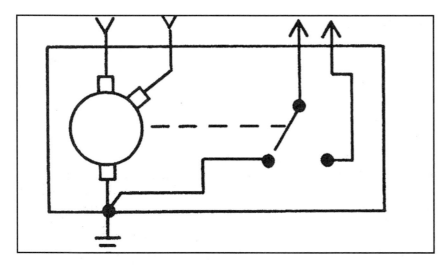

This is a typical way to diagram the wiper motor because it has several positions built in. Although it looks too simple for the real thing, it is quite accurate.

key on to test each wire, including the ground, for current when the switch is turned to each specific function position. These would include: wipers on low, medium, high and intermittent if they have this function. Just pull the connector from the motor and one-by-one test each with a test light or multimeter when the switch inside is turned to each position/function. You should have voltage at one or more connectors. This depends upon how the circuit and motor are wired. If you have current at all positions and a good ground the wiring is fine. If there is not +12 volts at each position you have a problem between the switch and the motor.

I would check on either side of the bulkhead connector, one wire at a time, for the required power at each switch position. If this checks out fine, and you have no power at the motor connector, you have an open wire between the firewall connector and the switch connector. Trace it back and fix it. But, do not stop there. I would also do a voltage drop of the motor in each function as that will tell you if the motor is fine, failed completely or possibly failed in just one speed or position.

Do not forget that the wiper arms and blades are just as critical to the windshield wiper system as is the motor itself. I change blades every year just before the bad weather. This year I also changed the wiper arms so I would have fresh springs pulling the blade tight against the window.

Ignition points were around for about 75 years, starting with some of the first vehicles built after the turn of the century to the mid-1970s. They lasted because they were functional and cheap, plus they generated additional service work on a regular basis. But they couldn't meet EPA emissions guidelines, which stated that a vehicle must stay in compliance for a minimum of 50,000 miles from the day it was purchased. Points needed to be changed about every 15,000 miles to keep an engine running properly. Therefore, the points ignition system was replaced with electronic systems in new cars.

Chapter 7

Auxiliary Electrical Systems

This chapter contains five sections: ignition systems, electronic fuel injection, air conditioning, power windows and door locks, and auxiliary driving lights. How important these are depends on whether or not these items are being installed on your project. Electric fuel pumps and high-performance fuel systems are essential for the operation of any vehicle with electronic fuel injection regardless of expected power output, but not if your engine is carbureted. Power windows and A/C might be going into your street rod, but not your race car.

If you are driving a classic car you will want to know how to get rid of points that require a tune-up every 15,000 miles. Driving lights are covered here as well. If you are driving off-road or touring the redwoods working on a lot of curved roads after dark, they are important. But if you are in a major city and never drive secondary roads, not too important.

ELECTRONIC IGNITION SYSTEMS

Electronic ignition systems were first mandated as standard equipment in 1974 because of the 50,000-mile emission durability test required by the EPA. The problem with the old system, which had been used for nearly 75 years, was the points started to deteriorate after 1,000 miles, and were totally worn out by 20,000 miles. Electronic ignition systems and computers on vehicles are used to produce hotter sparks

at exactly the moment they are needed, so they are much more efficient. Spark timing is calibrated by the ECU, which processes data from the engine sensors making precise decisions based on the engineering requirements for the engine. Unless you are restoring a vehicle to original condition for show competition, you will want to replace an old points-style ignition system with a more modern one. If you want it to look stock there are kits available to do this, but with much better performance and economy than a set of points.

Mechanical vs. Electronic Ignition Systems

The basic function of the ignition system has changed little over the years, but innovations in the last few decades have made spark timing much more precise. The basic components of the ignition system are the ignition coil, high-voltage ignition wires, distributor and, of course, the spark plugs. Your 12-volt battery is the source that delivers voltage to the spark plugs to ignite the mixture at the proper microsecond in the cylinder. But a 12-volt system does not deliver enough voltage to create that spark in the electrode, which requires thousands of volts. Where do we get these thousands of volts at just the right time? The ignition coil (a transformer) is the key. The coil has two sides. The first is the 12-volt or primary side. (This side has a few hundred turns of a large diameter wire and builds up the magnetic

After removing the distributor cap (with wires), leave it in the car, or remove the entire distributor as shown here (easier in a vice than in the car). In either case remove the rotor and this is what you will see. Points, condenser (actually a capacitor), the mounting plate and the advance mechanism, which are the two weights and springs on top.

This is the M&H Electric Fabricators "Breakerless Single Wire" point-replacement kit for all GM 1957 to 1974 plus AMC/Rambler V-8 distributors. It includes everything you will need including excellent instructions. The parts in the kit are the two halves of the firing ring (top left), the electronic module (top right), hex wrench and small screwdriver tip (bottom left) and all fasteners (bottom right).

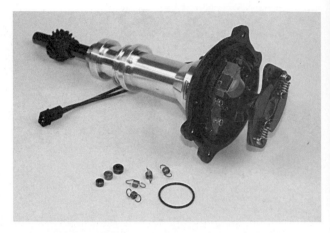

This photo of an MSD billet distributor is included so you will see that the stock distributor and the electronic one are very similar, with the main differences being the billet body and higher-quality components. This distributor can be installed and never touched for 100,000 miles or more. If you do not have to leave the stock distributor in your vehicle to maintain OE equipment, this type of new distributor would add style and performance to any hot rod.

field in the coils.) The other side is the high voltage or secondary side. (That side has thousands of turns of smaller diameter wire.) The coil uses "electromagnetic induction" to boost or transform 12V to the required high voltage. When we turn off the voltage on the primary side, the collapsing magnetic field induces a voltage in the secondary side, producing the thousands of volts the spark plugs need. In a traditional ignition system, the switching on and off of the primary voltage is done with a set of breaker points. The points are set inside the distributor and ride on a cam on the distributor shaft. This cam will have one lobe for each cylinder. When the points were closed, current flowed into the primary side of the coil creating the magnetic field. When the cam lobe opens the points, the current is turned off and the magnetic field collapses, sending a high-voltage peak to the spark plug wire that delivers it to the spark plug where it jumps between the center electrode to the outside wire that is part of the body of the plug. Of course, this only works if the plug is grounded to the head and engine.

Timing—Ignition systems also have an ignition timing variation called advance/retard to increase fuel economy. Timing is adjusted from weights in the distributor or by using the vacuum from the

inlet manifold or a combination of both. This function is replaced on fully electronic systems with a more sophisticated electronic control. But most of the high-performance distributor systems continue to rely on both mechanical and vacuum controls. An electronic ignition system uses a transistor to turn on and off primary power instead of points. Transistors are electronic switches that either work or don't, they don't deteriorate in use.

Electronic systems are capable of producing up to 45,000 volts and much higher amounts of current than the breaker point system. The points are

In this close-up showing both the OE (left) and high-tech billet distributor (right), you can see the difference in the area below the cap. The points and condenser are visible in the OE distributor and the control module can be seen on the MSD on the right. Inside the MSD are bearings not bushings, a quality feature that the factories never used. The advance mechanism on top works exactly the same but the stock unit has stamped weights and the other has forged weights.

This modern computer (ECM) controls everything in the vehicle's powertrain as well as other functions. It replaces the distributor and advance mechanism for the ignition system. It can be inside the passenger compartment or outside in the engine bay. This Delco unit is for a Chevy pickup truck and its mounting plate protects the entire bottom of the ECU from moisture and foreign objects.

The first step to changing over to a no-points distributor is to remove the old points by removing the hold-down screws. You will then loosen the screw holding the two wires to the point assembly, and remove the two screws holding in the point assembly. Then remove it from the distributor. Some kits have no screw holding the wires on, just a snap fit. Then remove the two wires on the points (screw or snap). Make sure the ground wire below the point plate is in good shape or the system could fail. Also take the condenser out and toss it.

replaced with a control module and the lobes on the cam are replaced with a trigger device. The trigger device uses a magnetic field to induce a small "trigger" voltage in the control module to turn off the current to the coil. As it passes, the module turns the current back on. It is extremely accurate, can be used to produce higher voltages and does not need maintenance. These systems can be bought off the shelf to replace the points system on classic cars so that the distributor cap looks standard and all the electronics are contained within it.

ECUs and Electronic Ignition Systems

In a modern vehicle with an ECM (electonic control module) unit and computer-controlled ignition system, the control module and the triggering device are replaced by a crank angle sensor (CAS) and an ignition control unit. Now engines no longer need a distributor. The CAS has a plate that has 360 one-degree marks, four 90-degree marks and two 180-degree marks. There is an infrared sensor that "sees" these marks and tells the control unit exactly where the crankshaft is and the control unit turns off the current to the coil at the precise instant the spark is needed. Since the control unit can do many calculations per second and the CAS doesn't have to be in the distributor, manufacturers simply did away with the distributor entirely. Now a separate coil is provided for each spark plug.

The ECM signals each coil independently. Since the CAS tells the ECM where the crankshaft is at

any given moment, it's easy for it to fire the plugs at the optimal moment. These changes have made the system more reliable, but have also allowed the engine designers more latitude on the timing. The spark will be fired at different times in relation to the crank position depending on engine speed, whether it is under load and other factors.

In conventional and electronic systems if the coil went bad, you were stuck. Now with 4, 6 or 8 coils, losing one coil will affect engine performance, but you can still drive. Speaking of reliability, most engines will continue to run even if the CAS fails. The ignition control unit stores basic timing data and when the signal from the CAS stops, it will go into a "fail safe" mode and use the basic data to continue engine operation. In most cases the fail-safe mode will limit speed and/or engine rpm to protect the engine and the "check engine" malfunction indicator light will come on.

Points Replacement Kits

As explained earlier, points are an old solution to an ignition and not a very good one. They work for a while and then need to be replaced or cleaned and reset. In today's world there is no longer any need to use points in any engine, because electronic systems are able to replace the function and eliminate the side effects. We will now show you how to make the change from points to an electronic ignition using older point-type distributors built from 1957 through 1974. You can do the installation in the vehicle or you can remove the distributor from the vehicle. Just make sure you do not drop any screws or washers into the body below the point plate. Mark the position of the rotor in relation to the engine block using something that will not rub off while you are working on it.

By using one of the replacement kit systems on

Everything needed for the conversion is contained in the M&H package, along with a very good set of instructions. This kit replaces the original points in early GM distributors. Since there is only one wire, it still looks like the original and no one can tell unless they take off the distributor cap.

Take the two ring halves and put them together. Take one of the two long bolts and start it from the bottom up (as shown) and install the assembly, open in the middle from the bottom of the mechanical advance assembly up into one of the two rotor screw locations. Do not use a lock washer or anything else under the head of the bolt.

This is what the distributor should look like with the two halves of the firing locators installed correctly.

You will need to use a small screwdriver inserted through the screw hole from the top to turn the bolt counter clockwise and draw it up into the hole until it bottoms out. Do the same with the other bolt, closing the circle around the distributor cam.

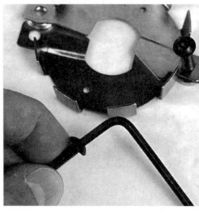

Once they are both installed and bottomed out, use the supplied hex wrench from the bottom and tighten both bolts tight. Make sure to tighten both bolts equally and make sure the ring halves are aligned and solid.

the points plate. These are harder to modify but just as stable a platform as the GM units.

GM Kits—GM distributors have an adjustment window (1957–1973) in the side of the cap that allows the points to be adjusted externally with the engine running. It is done with a dwell meter and a hex wrench. It is a simple procedure, much easier than setting the points with a feeler gauge, but no more accurate when done by a competent mechanic. (GM distributors have been mounted in the back of all V-8s since they were introduced in 1949.) These distributors have their centrifugal, mechanical advance system above the contact mounting plate in the top of the distributor, right below the rotor, great for easy maintenance and performance modifications.

The change to the HEI ignition with the coil in the cap of the distributor started in late 1974, thus ending the window cap's long 17-year run. Although it was a big improvement, it took the aftermarket a while to design improvements to it for higher rpm and more power. Today this design is still used when hot rodding older/non-computer engines for street applications as well as racing.

M&H Electronic Fabricators developed a single-wire design for 1957 through 1974 GM and AMC/Rambler V-8 distributors. Sales of this product are growing because it looks stock with just one wire and no one can tell the difference without removing the distributor cap.

To install, begin by removing the points and condenser and set them aside. The first thing to install is the split-firing ring that gets mounted under the mechanical advance mechanism. Start by installing one of the two hex head screws from the bottom of the split ring assembly (both pieces). Do not use any lock washers on it. Take the distributor and start the long screw from the bottom of the advance assembly up through the rotor screw location. Since it is coming from the bottom, start it with your fingers, or use the recommended small screwdriver. Once you have one screw installed,

the market, you can easily make the change and reap the benefits with smooth idle, more power and no maintenance. They all fit inside the distributor, do not need anything that doesn't look stock and they deliver outstanding performance for the life of the installation. In the performance V-8 category, there are two basic types of distributors used. The first is was used by all General Motors and AMC/Rambler V-8s from 1957 through 1974 and they have their advance mechanism above the points plate. The second distributor was used by Chrysler and Ford from 1955 to 1974 that has the centrifugal/mechanical advance mechanism below

Slip the new firing module into this space where the points were located. Use the new mounting screws and make sure all wires are securely mounted and not under the firing module before you lock it down. It is best to install the wire to the coil, already inside the distributor housing, before slipping in the module.

You still have to orient the rotor so the square peg slides into the square hole and the round peg does the same on the other side. The two mounting holes in the rotor usually need to be enlarged to allow the rotor to easily slide over the vane mounting screws. A 7/32" drill, small file, or plumber's ream can be used to open up the holes slightly.

IMPORTANT: The square and round indexing pegs on the bottom of the rotor must be shorter than 1/8" for the rotor to seat properly. The 1/8" hex wrench can be used as a thickness gauge or you can also use your digital dial calipers.

This is all that is in the conversion kit, an electronic module with two wires coming off it and the firing ring. The small round ring fits over the cams on the shaft and does the same job as the split-firing ring in the M&H kit.

If you want to go fully electronic as well as high tech, this MSD Chevrolet billet distributor is just the ticket. It even retains the vacuum advance mechanism.

close the split ring and install the other. Once this is done, tighten both using the small hex key wrench.

You are now ready to install the module that controls the firing electronically. Next install the black electronic magnetic pickup module, mounting it to where the points were located before. You will need to turn the firing ring to an open space to get it into position. Once the two mounting screws with lock washers are installed you should be able to spin the top of the distributor and the firing ring tabs (eight of them) should slip through the slot in the magnetic pickup.

Next, turn over the rotor and file it down until about 1/8" to 3/32" remains. Now you can install the rotor (new is best) and tighten it into place using the lock washers and nuts in the kit. That is all you do inside the distributor.

If you removed the distributor to do the installation, just install it with the rotor pointing to the same position as when it was removed. Install the single wire from the coil using the black washer/wire clip under the negative side of the coil and you are good to go. At this point you should

have changed the cap to a new one and be ready to install it with the existing or new plug wires. That's all there is to installing the M&H single wire point kit in an early GM distributor. Easy as 1, 2, and 3, and except for changing the plug wires and cap (if they are old), should take less than one hour total.

Ford/Chrysler Kits—Chrysler and Ford vehicles used a different type of distributor from 1955 to 1974. One significant difference is that the centrifugal/mechanical advance is located below the contact mounting plate. This difference is why there are two designs in point replacement kits. All of the other kits, even for GM, require two wires to function. That doesn't make them bad—it just distinguishes them from the M&H unit.

Pertronix, Crane and Mallory offer a two-wire system. Since all OEM points systems had used a single-wire design, it takes a bit of work to hide the second wire but by taping them together it can be done. With either type, overall performance is

This is an original distributor from a 1969 Mustang that is being rebuilt and upgraded. After installation, we will cover the red wire on the outside of the housing with shrink tubing plus tape and it will be almost invisible. For major shows the owner installs an original 1969 Mustang distributor for restoration competition. But for everyday driving or weekend cruising, the Pertronix points replacement module and Flame Thrower coil are hard to beat.

The ground wire needs to be inspected for breaks or damage. This one is old, but it will work just fine.

To install the module you simply slip it in place where the points are located and it will settle into where it belongs. Next, take the ground wire and use the mounting screw (the same one that held in the points) slip it through the ring connector and screw it into place. There should be plenty of clearance between the firing ring and the module, so it can turn easily. I prefer to leave it slightly loose and not tighten the screw down until the firing ring is in place.

improved and once set, remains set regardless of mileage. Installation is actually simple with either type of system. Pertronix, Prestolite and Ford and Chrysler installations are essentially the same for all distributors with the mechanical advance below the contact mounting plate.

On any Ford engine with the distributor in front, it is probably easiest to leave the distributor in place when making the changeover to the Pertronix system. Just make sure you do not drop any screws or washers inside the body of the mechanism. Start by removing the distributor cap, and set it aside with the spark plug wires still attached. Since I will remove the distributor to make photography easier, I marked the position of the rotor on the engine as well as inside the distributor with something that will not get rubbed away. (I use a little correction fluid.) Next, remove the rotor and set it aside as you will be using it again. If it is burned or pitted, this is the time to change it for a new one and a new cap as well. Remove the points and condenser and discard; they are not going to be used again.

Disconnect the wire that is connected to the coil and pull it back through the distributor housing. The grommet will also be removed at this time. Take the module and install it where the points had been located. It will fit only one way so be patient and it will slip right in.

Once it is mounted, slide the two wires (red and black) through the side of the distributor housing from the inside to the outside and run them to the coil. In the process pull the grommet until it seats

into the housing. Make sure the wires do not interfere with moving parts. It is important to make sure the ground wire inside the housing is in good condition and connected.

If you have taken the distributor out for these modifications, sliding the distributor back into the engine can require a bit of fiddling. As you slide it back into the engine, aim the rotor for the marks both inside as well as outside the engine, but start it a little short of the position you want it. Since the gears are cut on a curve (helical) it will cause the rotor to move a bit at the last as it slides into place. If you don't get it right the first time, pull it up a bit and move the rotor a little in the direction you want it before you slide it home again. Once it is in and aligned correctly, you are ready to snap the cap and wires back into place .

With the Pertronix system, you have a black and red wire coming out of the distributor. The black wire goes to the negative post of the coil and the red wire goes to the positive post. If you want to hide the red wire, tape the two together with black electrical tape or shrink tubing and the red wire will be almost invisible. Although it will appear factory-stock on the outside, you now have an electronic ignition without points.

There are many performance aftermarket ignition boxes, distributors and wires available that can improve the ignition by a large margin; it depends a

These three photographs show how to insert and pull through the module wires from inside to outside the distributor housing. Pick up the two wires from the module and slide them and the grommet from the inside to the outside of the distributor body (left). Doing it the other way has been done, even by experts, but if you do, you will just have to pull it back out and this time start the wires from the inside. Make sure the wires are pulled all the way through and that the grommet slides in place (middle). You can see the grommet for the two wires as they get pulled through the hole from the inside. Keep pulling gently until all the extra wire is through the hole but not tight and the grommet is in place sealing the distributor from moisture and water (right).

The firing ring cannot really be set on out of position. Turn it a little at a time and use a slight downward pressure until you feel a slight drop; push it all the way down gently. Now mark the ring 180 degrees from the rotor slot. This way when you go to install the distributor again, you will know where the number 1 plug wire is.

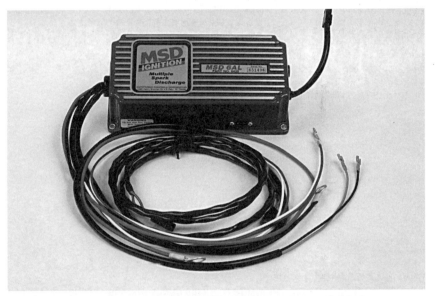

Pertronix, MSD, M&H and others offer high-performance coils to match their ignition systems. They are available in stock black or chrome. They can deliver more than 40,000 volts to the plugs.

Aftermarket high-performance ignition systems are also available from several manufacturers, but the company with the most experience in the field has to be MSD, who has at least two duplicate ignitions in every NASCAR-series race vehicle (cars and trucks alike). Their 6-AL spark ignition delivers multiple sparks at low rpm for drivability and also allows rev-limit chips to protect the engine from over-revving. This ignition box, as well as others from MSD, can actually be fired with points, but since the points would be carrying almost no voltage they would last at least 100-times as long. Still, if you want reliable high performance, splurge for MSD's high-performance billet aluminum electronic distributor.

lot on the application. M&H and Pertronix ignitions offer a stock, replacement look, but systems like those from MSD, Mallory and Crane are designed for maximum-performance motors. If you are drag racing your street car, have a specialized drag car or do any other form of high-performance driving, you'll need a performance aftermarket unit.

As you can see, this 351 Ford MSD billet distributor has a polished aluminum body and a unique top construction. It is similar to Ford electronic distributors in some ways, but because of the GM HEI-style cap lugs, the spark plug wires stay on because they can be locked on with the locking plate and two plastic screws that fit into the cap. This very unique system is a real benefit.

This fuel pump assembly is typical of high-performance aftermarket electric fuel pumps. Although this is for more than just a stock grocery getter, it is basically no different from a stock fuel pump system. The major difference is that this pump can flow three times the amount of the stock pump and it is not mounted inside the tank. To use this system you will need to run a larger fuel line to the front of the vehicle and then use the old line to return fuel to the tank.

In many applications it is best to have the ignition box and coil inside the vehicle to protect them from the elements. To facilitate this, MSD makes this bulkhead connector so you can run the high-voltage coil wire through the firewall without any risk of the voltage shorting out on the way. Regardless of what pointless ignition you select, just remember— "no points" will deliver better performance, longer spark plug life and better fuel economy. You can keep it cheap or go high tech, which is simply a matter of taste and actual performance needs.

This coil is matched to the MSD 6-AL system as well as MSD's billet distributor. Just follow the instructions and install. It will also work without the MSD 6-AL, with just the coil and distributor. Always match the coil to the ignition, otherwise you will be disappointed.

ELECTRIC FUEL PUMPS

Although electric fuel pumps themselves are easy to understand, the circuits that control them are a little more complex. All pumps have a fuse in the circuit, not always in the fuse panel, but more control is needed than just overload protection. This is especially important when controlling a highly flammable liquid—like gasoline. If the engine isn't running you do not want the electric fuel pump to keep running. Most pumps are powered during cranking and will continue to run after the engine starts. If the ignition is turned on without starting the engine, the pump only runs for a second or two. This safety measure prevents fuel from being pumped into the engine if an injector was leaking. It also prevents fuel from being spewed all over in case the vehicle is in an accident.

Before you go further into the system you need to make sure the fuel pump gets electricity during start and when running. Some of the vehicles you will run into have unique protection to make sure the fuel pump cannot run when the engine is not running. A computer-controlled relay is the most common method of controlling the pump. The computer looks for an rpm signal to decide if the engine is running. If it sees the proper signal, it grounds the relay, and, in turn, powers the pump. You will find slight variations among control circuits. Regardless of the system, the fuel pump must receive power when the key is in the run position. This is easy to identify with a test light.

Ford circuits, for example, have an inertia switch

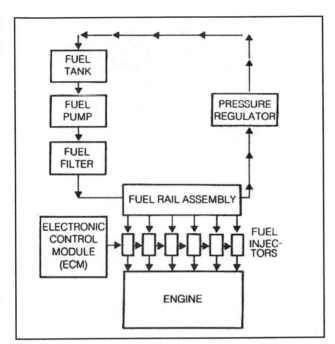

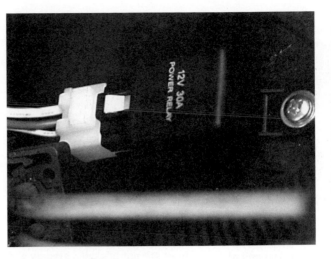

This diagram shows how the fuel-delivery system works on a fuel-injected vehicle. In this case it shows direct port fuel injection on a 6-cylinder vehicle. If you add two more cylinders for a V-8 or take away two for a 4-cylinder, everything remains basically the same.

This is a typical aftermarket relay and could be used for an electric fuel pump on an aftermarket EFI system. OEM or aftermarket, they all do the same job as any other relay; they activate a high-current component using lower current to make it switch on.

This Hobbs's switch goes to ground when it has pressure which completes a circuit and lets it operate. This is set at 50 psi, but a fuel-system switch as used by GM is set at less than 10 psi oil pressure.

Relays can be placed anywhere, but normally you will find them all together if it is an OEM vehicle system, but aftermarket systems might be by themselves. This row of OEM relays all use the same part number, so check your vehicle wiring diagram or go online to know which one is the correct relay. Or get out your multimeter and see which one gets power when the ignition switch is in the run position.

If the fuel system includes a carburetor, all you have to do is pump the throttle once by hand and you will hear the fuel squirting into the intake manifold.

that disables the pump in the event of an accident. This switch is a lot like the switches that operate air bags on vehicles. If there is a high load of inertia imposed from any angle, the switch opens the circuit so no fuel is pumped. This is an accident safety system that works well, but may be a bit more costly than the GM system.

GM's safety control circuit has a relay bypass that powers the pump through a pressure switch in case the fuel pump relay fails. If an engine is running it is building oil pressure. A pressure switch is a

If you are using a TBI system, you will see the fuel pulsing into the throttle body when the engine turns over. But, if it is a direct-port system you need to hold your hand on one of the hoses going to the fuel rail. If it is working you will feel the hose pulse.

This is the fuel pump access panel (cover removed) on an Acura hatchback. Remove the back seat and you can get right at it. Most vehicles are not this user friendly and you will have to remove the tank to get at the internal pump.

It doesn't matter if the ECM is an aftermarket unit or OEM, if they fail they are expensive. The good thing is that they seldom fail. Check everything else before you start blaming the ECM for a lack of fuel pressure. One way to test the pump is by wiring it directly using a jumper wire and a ground wire. If it has 12 volts to the power-in side and a ground, it should run.

Here you can see the type of stock OEM tank access panel. This is on a modified 1965 Mustang. The fuel outlet pipe has been covered with a rubber cap and a sump has been welded to the bottom of the tank. This assures that there will be plenty of fuel for the modified engine. The stock tank access panel looks just like the ones from OE's when there is a fuel pump inside. Since they are for electronic fuel injection systems they will also have another smaller pipe going back into the cap as a fuel return. Note that the fuel line has been hit hard by a rock or other road debris, so it needs to be changed.

Using a multimeter or a jumper cable, ground the fuel pump (by taking the wire off of the oil pressure safety switch) and have someone turn on the vehicle to the start position. If you hear the fuel pump working, the pressure switch is bad. Replace it and try to start the vehicle.

simple and very effective method to prevent a fuel pump from operating after the engine is shut down or has quit running. If a vehicle was on its side or upside down it would not have any oil pressure either. It is just in the start position that the circuit is closed before the engine is running. The fuel pump relay is located in the engine compartment. Other than checking for loose electrical connections, the only service necessary is to replace the relay if found defective. A voltage drop test will let you know if it is good. If it receives power and doesn't make the bigger circuit "hot" it is a failed unit. If the oil

pressure switch is bad, replace the unit.

When you need to work on the fuel tank, make sure there is no fuel in it by rocking the vehicle side to side and listen. You can also run a plastic hose down the filler neck to see if there is fuel in the tank. Just remember, fuel tanks are dangerous, even more so when empty, so be careful when you remove it from the vehicle. On some vehicles there is an access plate below the rear seat that will allow you to test the circuit and replace the pump, if required, without removing the tank itself. Vehicles with this construction are very cool and easy to work on.

The electronic control module (ECM) is sometimes referred to as the Engine Control Module. It accepts data from various sensors and switches to calculate the amount of air and fuel needed to achieve the ideal stoichiometric ratio of 14.7:1, air to fuel, in order to optimize catalytic converter operation and reduce emissions as much as possible. If a system failure occurs that is not serious enough to stop the engine, the ECM will illuminate the "service engine soon" or "check engine" light and will continue to operate the engine, although it may need to operate in a backup or fail-safe mode. Fuel is supplied to the injectors by an electric fuel pump assembly which is usually mounted in the vehicle's fuel tank. The ECM provides a signal to operate the fuel pump through the fuel pump relay and oil pressure switch. If the pump motor quits, the vehicle will not run.

Other system components include a pressure regulator, an idle air control (IAC) valve, a throttle

On most TBI systems, the fuel pressure regulator is easily found on the throttle body assembly as seen here on the Holley TBI unit. It is between the two hose nipples and is adjustable using a hex wrench. It is best to leave it as set by the manufacturer unless one of their technical people recommends you adjust it either up or down.

By the early 1960s, 25% of all American cars were equipped with air conditioning. Today it is standard equipment on nearly every car; you have to special order to get one without it. Classic Mustangs like this one were factory equipped with A/C, but only as an option. Many enthusiasts of sports cars didn't want the A/C pump to siphon off precious horsepower, and to a small extent, they were correct.

Here you can see a complete aftermarket high-performance electric fuel pump installation for an EFI system. The fuel comes from the sump to the inlet side and the #1 fuel filter. Then it goes into the pump and it goes back out and to the front of the vehicle after going through another filter.

position sensor (TPS), manifold air temperature (MAT) sensor, coolant temperature sensor (CTS), a manifold absolute pressure (MAP) sensor and an oxygen sensor. The fuel injectors are solenoid valves that the ECM pulses on and off many times per second to promote proper fuel atomization. The pulse width determines how long an injector is on each cycle and this regulates the amount of fuel supplied to the engine.

The system pressure regulator is part of the throttle body fuel meter cover which is designed to keep fuel pressure constant at the injector regardless of engine rpm. This is accomplished by controlling the flow in the return line (a calibrated bypass).

AIR CONDITIONING

Nearly all late-model vehicles come with air conditioning as standard equipment. We've become so accustomed to it, that enthusiasts are now installing retrokits like those from Vintage Air into street rods and muscle cars. So how does it work? What should we do to keep it working at peak efficiency? What should you look for if you are considering adding it to an old vehicle?

Air conditioning was first introduced in vehicles in the 1940s. There have been many improvements since then, including better design, new cooling gases and of course, the electronic climate control systems. Now the whole system will automatically maintain the car to a preset temperature, and many offer dual climate where the passenger and driver can each set their own temperature.

Refrigerant

The cooling liquid used in older vehicles used to be Freon, the tradename for R-12 once manufactured by DuPont. In the 1980s, it was banned from use, so any older vehicle that needed to be "recharged," had to use R-134a, the more environmentally friendly replacement. But this is not so easily done on an older car, and you can't do it yourself. You'll need a shop that is trained and certified to purchase and handle the R-134a refrigerant, but a little knowledge will help you make informed decisions.

This new refrigerant has a higher operating pressure, therefore, your system, dependant on age, may require larger or stronger parts. This will add significantly more cost to the final repair. And if not performed properly, may reduce cooling efficiency which equates to higher operating costs and reduced comfort.

This is all you will generally see of your A/C system; the controls that make it go to work when you want to. But, there are two moving components and several other critical components that allow it to do its job.

A/C System Components

The most common components which make up these automotive systems are as follows:
- the compressor which provides the power
- the condenser which cools the refrigerant
- the orifice tube or thermal expansion valve to keep the temperature of the coolant even as the liquid changes to a gas
- the evaporator which working with the blower motor sends very cold air into the cabin
- the receiver/dryer or accumulator which protects the system
- the refrigerant that, when under pressure, becomes a liquid. Traditional Freon is now illegal and has been replaced by R-134a, which has different properties and cannot be used as a direct replacement without changes to the system.

Compressor—The compressor is a pump, belt-driven by the engine, which is the heart of the system and provides its power—the power of several fridges and needing several horses of dedicated engine power. A big system for a luxury vehicle or a big van could draw as much as 45 hp when the system is on. But when it is off, the power drain from parasitic losses is minimal. The A/C system is split into two sides, a high-pressure side and a low-pressure side; defined as discharge and suction. Since the compressor is basically a pump, it must have an intake side and a discharge side. The intake draws in refrigerant gas from the outlet of the evaporator where it has absorbed the heat from the air inside the vehicle. In some cases it does this via an accumulator. It is compressed and sent in liquid form to the Condenser where it transfers the heat absorbed from the inside of the vehicle to the ambient air.

Condenser—The condenser is really just a big radiator to draw off some of the heat in the refrigerant caused by it being compressed. It pulls as much heat away as possible using ambient air. From there it gets sent to the evaporator.

Evaporator—Located inside the vehicle, the evaporator cools the air for driver and passengers alike. The evaporator provides several functions. Its primary duty is to remove heat from the inside of your vehicle by letting the blower fan motor send hot air from inside the cabin through it. While doing so, because of the laws of physics, it removes heat and moisture, acting as a dehumidifier. As warmer air travels through the aluminum fins of the cooler evaporator coil, the moisture contained in the air condenses on its surface. Dust and pollen passing through stick to its wet surfaces and drain off to the outside. On humid days you may have seen this as water dripping from the bottom of your

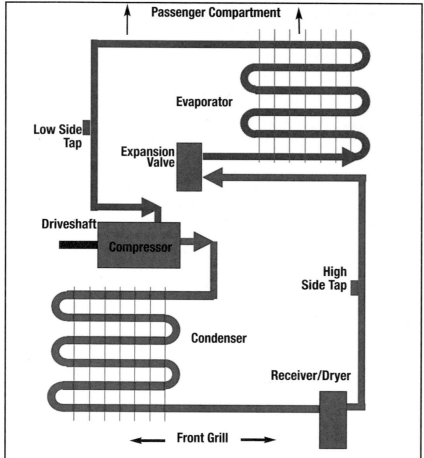

Here is a schematic of a basic A/C system. Note that the compressor draws in R-134a refrigerant from the low-side tap and compresses it until it reverts back to a liquid. Next the refrigerant is in the high side and is sent to the condenser at the front of the car (usually behind the grille) that allows outside air to extract the heat built up in the liquid from being compressed. Next it is sent to the expansion valve under high pressure where it then goes to the evaporator. This creates very cold R-134a, which is then passed through the evaporator located inside the vehicle, which is the start of the low-pressure side of the system. The A/C blower motor blows cabin air through the evaporator under the dash and what comes out is cold air to make the cabin comfortable for driver and passengers alike.

The compressor of the system does just what its name implies. It compresses R-134a in a gaseous state and turns it into a liquid state. This is the beginning of the high-pressure side of the system. It has two working components, the electric clutch and the compressor itself.

vehicle. Rest assured this is perfectly normal.

The ideal temperature of the evaporator is 32°F or 0°C. Refrigerant enters the bottom of the evaporator as a low-pressure liquid. The warm air passing through the evaporator fins causes the refrigerant to boil (refrigerants have very low boiling points). As the refrigerant begins to boil, it absorbs large amounts of heat. This heat is then carried off with the refrigerant to the outside of the vehicle.

Several other components work in conjunction with the evaporator. As mentioned above, the ideal temperature for an evaporator coil is 32°F. Temperature and pressure-regulating devices may be used to control its temperature. While there are many variations of devices used, their main functions are the same; keeping pressure in the evaporator low and keeping the evaporator from freezing. A frozen evaporator coil will not absorb as much heat, and will block the drain, possibly allowing the water to stay inside the vehicle on the cabin floor after the ice melts.

Pressure Regulating Devices—Controlling the evaporator temperature can be accomplished by controlling refrigerant pressure and flow into the evaporator. There have been many variations of pressure regulators introduced since the 1940s. The most common is the expansion valve or orifice tube used by Ford and GM. Chrysler tended to use a variation called a thermal expansion valve.

Expansion Valve—This orifice tube, probably the most commonly used, can be found in most GM and Ford models. It is located in the inlet tube of the evaporator, or in the liquid line, somewhere between the outlet of the condenser and the inlet of

the evaporator. You can find it by locating the area between the outlet of the condenser and the inlet of the evaporator where there is a sudden change from hot to cold. You should then see small dimples placed in the line that keep the orifice tube from moving. Most of the orifice tubes in use today measure approximately three inches in length and consist of a small brass tube, surrounded by plastic, and covered with a filter screen at each end. It is not uncommon for these tubes to become clogged with small debris.

While inexpensive, the labor to replace the valve involves recovering the refrigerant, opening the system up, replacing the orifice tube, evacuating and then recharging. With this in mind, it might make sense to install a larger pre-filter in front of the orifice tube to minimize the risk of this problem recurring. Some Ford models have a permanently affixed orifice tube in the liquid line. These can be cut out and replaced with a combination filter/orifice assembly.

Thermal Expansion Valve—A common refrigerant regulator is the thermal expansion valve, or TXV. It is commonly used on import and aftermarket systems. This type of valve can sense both temperature and pressure, and is very efficient at regulating refrigerant flow to the evaporator. Several variations of this valve are commonly found. Another example of a thermal expansion valve is Chrysler's "H block" type. This type of valve is usually located at the firewall, between the evaporator inlet and outlet tubes and the liquid and suction lines. These types of valves, although efficient, have some disadvantages over orifice tube systems. Like orifice tubes these valves can become clogged with debris, but also have small moving parts that may stick and malfunction due to corrosion.

Receiver/Dryer—The receiver/dryer is used on the high-pressure side of systems using a thermal expansion valve. This type of metering valve regulates liquid refrigerant. To ensure that the valve gets liquid refrigerant, a receiver is used. The primary function of the receiver/dryer is to separate gas and liquid. The secondary purposes are to remove moisture and to filter out dirt. The receiver/dryer usually has a sight glass in the top. This sight glass is often used when charging the system. Under normal operating conditions, vapor bubbles should not be visible in the sight glass. When checking a system using R-134a systems note that cloudiness and oil that has separated from the refrigerant can be mistaken for bubbles. This type of mistake can lead to a dangerous overcharged condition. There are variations of receiver/dryers and several different desiccant materials are in use.

This close-up of the compressor clutch not only identifies it from the alternator, but it gives you an easy first step in troubleshooting the system electrics. The compressor clutch locks the drive belt to the compressor making it turn when it is on. It has two wires going into it, power and a ground. If there is no power at the clutch look at the system on/off switch. It could be busted, have a bad ground or of course, a blown system fuse. Have someone turn the A/C switch on with the key at run and you should be able to hear the clutch click on and off. If not, test the clutch lead with the A/C switch turned on and the key in the run position. There should be 12 volts to it and the ground should complete the circuit. To make sure the clutch isn't slipping, have the engine started up so you can look at the clutch as the driver slowly raises the rpm and turns on the switch. It should click and the motor should lose about 150 to 200 rpm under the A/C load.

On this vehicle, the A/C system turns on by moving the switch to any of three positions. This tells the system how much cold the driver wants, and where to send it. Green/blue is cold and red is hot regarding the temperature control knob. It also should engage the compressor clutch in all three positions. The fan switch just starts the fan blowing at the speed the driver selects. It is the same fan as used when the heater is turned on.

Some of the moisture-removing desiccants are not compatible with R-134a. The desiccant type is usually identified on a sticker that is affixed to the receiver/dryer. Newer receiver/dryers use desiccant type XH-7 and are compatible with both R-12 and R-134a refrigerants.

Accumulator—Accumulators are used on systems that use an orifice tube to meter refrigerants into the evaporator. It is connected directly to the evaporator outlet and stores excess liquid refrigerant. Introduction of liquid refrigerant into a compressor can do serious damage. Compressors are designed to compress gas, not liquid. The chief role of the accumulator is to isolate the compressor from any damaging liquid refrigerant. Accumulators, like receiver/dryers, also remove debris and moisture from the system. It is a good idea to replace the accumulator each time the system is opened up for major repair and anytime moisture and/or debris is of concern. Moisture is enemy number one for your A/C system. Moisture in a system mixes with refrigerant and forms a corrosive acid. When in doubt, it may be to your advantage to change the accumulator or receiver in your system. While this may be expensive, it will greatly extend the life of your air conditioning system.

Troubleshooting A/C Systems

Now that you know how it works in general terms, let's look at the problems it can have. We will not address the R-134a problems and the non-electrical parts of the system. They need to be checked by a qualified repair shop using special equipment. But, we will address the few electrical items that can fail.

The heater A/C blower motor switch should make the fan turn on and blow air through the system inside the vehicle (There should also be several speeds). If it doesn't, troubleshoot the fan switch which is usually right under the control panel in the dash. If it has power and a ground it will make the fan turn unless it is physically broken.

All other problems with the A/C will either be wires that you can trace, if the switch doesn't have power to it, or doesn't turn on from a bad ground. The possible electrical problems are few and finding and fixing them is easy. The real problems with this system are generally shorts to grounds and bad grounds. Then look at the switches. Not much else to worry about electrically. If you test these and cannot get the system to work, take it to a qualified repair shop and be ready to spend some money.

If you are working on one of the multi-zone A/C systems that controls temperature to the nth-degree, the complexity of the electrical elements go up significantly. Most of the temperature control issues are the temperature sensor and its communication to the control module for the A/C. (It might actually be independent of the ECM so check that out also.)

Aside from the logical things you can test with a

This control panel does the same job as the other, but it has a separate compressor switch (the A/C button) as well as a button to either continue letting outside air enter the cabin or stop the outside air and just cool the inside air, which will allow the system to work at maximum efficiency. Once again the temperature control knob has green/blue for cold and red for heat.

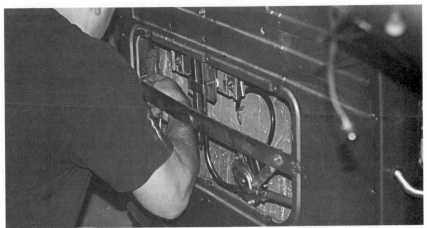

Steve Agnello is finishing up the installation of the driver's window on a 1956 Pro Street Ford pickup. You can see the lift mechanism and the wires. The whole process was easy from start to finish and probably the hardest step was adjusting the window and lift so it slid up and down without drag from the guides and moldings.

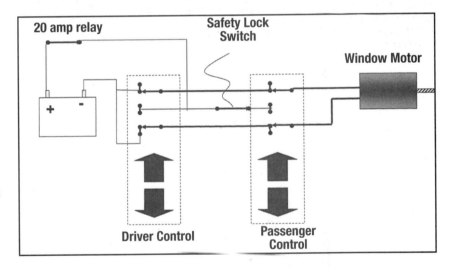

Starting at the battery positive terminal you can see that electricity flows first to a fuse and then to the safety lock switch. From there the system is ready to respond to any request from passengers as well as driver. In a real system the driver would have access to all window switches as well as the safety lock switch while passengers would only have control of their window. If the safety lock switch is engaged only the driver can open or close the windows. Note: This is a simplified system schematic, but all of the components are shown including the switches.

multimeter and visually, you will have to resign yourself to visiting your A/C professional shop for anything else. The good thing is that not much really happens to these systems during the first 5 to 10 years.

POWER DOORS AND WINDOWS

Most car owners have become so accustomed to the convenience of power-locking doors and electric windows, that they regard them as standard equipment. (When was the last time you hand-cranked a window?) Most enthusiasts now build their street rods with these systems in place, or retrofit them to classic cars.

Of course, power windows have been around for a long time, and the earlier systems were famous for ceasing to work or developing a mind of their own. That is mainly due to moisture getting into the door jamb and corroding components, or just old, weak motors that finally give out.

The major components of a typical system are the master control switch, individual window control switches, and the window lift drive motors. These include several yards of wire, but that is not a problem. The aftermarket has created kits to offer these features for restoring or building older vehicles. The electrical systems are essentially the same for a hot rod or a late-model car except for a few bells and whistles.

Electric Windows

Power electric windows are really a "must-install" item on any hot rod or street rod. Actually, it is

easier to install and wire an electric window opening and closing system than it is to replace or fix the stock hand-cranking window mechanism. It is probably less expensive also. When wiring power windows, you need to decide how many windows will be powered and where you want the switches. It is important for the driver to have full control over all powered windows in the vehicle.

With just two windows powered, the driver's door and the front passenger's, the easiest option is to place the switches in the center of the vehicle. Since

The same command module (7-channel) will be used for both the windows and the power-door opening system.

Here you can see two window switches right below the armrest. The passenger can control his or the driver can control both. The window safety lock switch was not needed in this vehicle. If it was a sedan the back windows could be added with just a little more work. The driver would then have five switches under the armrest. All four window controls plus a safety lock switch that would disable the passenger windows.

these vehicles will usually have bucket seats with a console, that is the most viable place to install them. Only two switches will be needed, since the driver will be able to reach them both. A safety-locking switch isn't needed in this application.

However, you may opt to place the switches in the doors and there are a number of reasons why this approach might be considered more practical. In this case you will need a third switch for the driver's door to mirror the passenger window switch. This option is popular when the front seat is a bench or for aesthetic reasons.

When fitting power to four windows, you have the same options but will need additional switches. You need two switches for the rear windows and they will each need a duplicate near the driver. Then the configuration of the front window switches will follow the same guidelines as for two. So you will fit either six or seven switches and all four windows will be controllable by the driver. There is an additional control that can be fitted to disable the switches (all passenger switches or just at the rear) which is useful when children will be in the vehicle.

Window Options—Next you need to consider what additional features you want. Options include one-touch switching and safety stop control. Finally, you need to consider whether you want the switches and electricity to run direct to the motors or through a relay. All options can add cost to the system, but these are excellent options.

• *Automatic-Down.* An automatic-down feature is fairly common on cars with power windows. You tap and release the down switch and the window goes all the way down. This is usually only found on the driver's window, although some higher-end cars have them for the passenger as well.

• *Automatic-Up.* The problem with automatic-up

windows is that if anything gets in the way of the window, such as a child, the window has to stop moving before it hurts the child. One way that carmakers control the force on the window is by designing a circuit that monitors the motor speed. If the speed slows, the circuit reverses the power to the motor so the window goes back down. On hot rods this would be costly, but possible.

• *Outside Window Control.* In this type of system the windows can be controlled from the key either while it is inserted in the lock or by using the remote control. Pressing lock for a few seconds will close all the windows and sun roof and unlock will open them all. This feature is very useful when the weather is hot.

• *Courtesy Power-On.* Some cars maintain the power to the window circuit after you turn your car off for a period of time, which saves you from having to stick your key back in the ignition if you forget to roll your window up. The power-window circuit will have a relay on the wire that provides the power. On some cars, the body controller keeps this relay closed for an extra minute or so. On other cars, it stays closed until you open a door.

For the investment in time and money, it is best to do the whole job while the vehicle is under construction instead of having to take it all apart later after wishing you had installed the extra features at the start. Get the most you can afford before you start. You can buy kits which meet all of these specifications.

Power Doors

When building a hot rod or custom car, it is the little things that separate a show winner from an "also ran." Electric door locks, shaved door handles and electric windows are three such features that provide that last bit of class to a show car. It is always fun to use a remote control to open the doors from outside a vehicle that has no door handles. Hit the button and the door pops open about 6 inches creating a bit of magic for the unsuspecting public.

Shaved Handles & Remote Door-Popper Systems—Door "poppers" and electric door latch releases are the secret to this magic. They are easy to install and can be operated by remote control with a pocket signal sender. The smooth look of a hot rod without door handles is beautiful. These systems have actually been around since the early 1940s using the same technology, except for the remote control electronics. Interestingly the system

The electric part of a door-popper system is very simple. But the part that opens the door six inches is nothing more than a spring inside a tube. Closing the door loads the spring that is inside the cab and touching the door on the bottom. Then, when the electric switch sends a signal to the door solenoid, it releases the door latch and the stored energy of the spring opens the door. This technology has been around since the 1940s. Of course there were no remote control systems then, but they are available today.

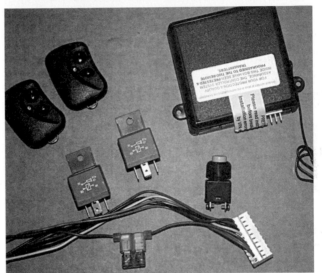

The complete remote control kit consists of the module, two remote control senders, two solenoids and two door-release manual buttons for a two-door vehicle. The wiring harness is not supplied with protective split tubing so add it yourself. The long black wire is actually the antenna for the system.

doesn't need an unlock feature because there are no door handles.

Door-popper systems use a solenoid to release the door latch and a spring-loaded tube to push open the doors. Inside you need to have two electric release buttons, one for each door. (If the vehicle is a four-door you could actually have all four doors operating off the system.) Do not place the inside release buttons on the center console as this could lead to accidentally opening doors if the switch button is pushed. On this vehicle they are located under the dash on each side next to the door. The driver and passenger must reach forward to activate them reducing the chance of accidental opening. There is also a mechanical way to open these doors in an emergency if there is no electrical power. Another method to keep the doors from opening while the vehicle is in motion is to wire the power through the park/neutral safety switch.

Shaved door handle kits use a control module that is generally 7 channels and a remote control to send a signal to open the doors. Once the button is pressed, it activates a solenoid inside the door. The solenoid is a magnetic-pulling device that can stroke 40 lbs. on each pull. The solenoid is connected to the door latch inside the door panel by a cable. When the signal is sent to the solenoid, the solenoid pulls a cable that pulls the latch open, which opens the doors. Installed in the door jam of the vehicle is a spring-loaded door popper.

Installing the shaved door handle/popper kit is

fairly straightforward. Begin by mounting the control module. The control module must be mounted away from any moisture, preferably under the dash tie strapped to a wiring harness or on the upper firewall using Velcro. Once the module is mounted, you will need to run your channel wires to each door. You will probably use the 7-channel control module. Channel 1 will activate to the driver's door, channel 2 the front passenger door, channel 3 and 4 for the rear doors (if any). Once you have the wires in the doors, you will need to attach your high powered solenoids to the door. Begin by removing the door panel. Inside your door, you should have a void where you can recess the solenoid. Use the straps and hardware provided to mount the solenoid. Keep it as close to and inline with the door latch as possible. Once the solenoid is mounted; connect the cable and crimp to the door latch then to the hole on the top of the solenoid. Then when the solenoid pulls, the latch pulls as well. When the door latch is pulled, there should be no bind. Now, connect the wires you have run to the positive side of the solenoid, usually

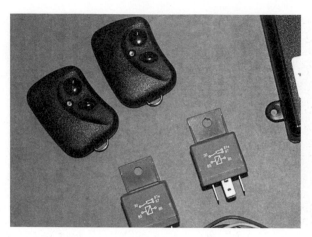

In this photo you can see the fairly large wires coming out of the spring tube with the door all the way open. The plastic grommet allows the spring to slide in the arc of the open to closed door. This is a simple solution to a difficult problem.

the white wire. Run the black wire into the cab of the vehicle and connect it to any chassis ground (–), or the negative battery terminal. Make sure you ground the solenoid inside the vehicle. Just grounding it in the door will surely cause malfunctions or intermittent operation sooner or later. That is why the ground should come back into the cab.

Once that is completed, you are left with a red, black and yellow wire on the control module. Connect the red wire to 12-volts constant (+), or the positive battery terminal. Connect the black wire (-) to chassis ground, or the negative battery terminal. The yellow wire must be connected to a 12-volt switched ignition feed. This wire must only have power when the vehicle is on. It will keep the shaved door kit remotes from working while the vehicle is driving. (Although all of this is what our system called out, the system you may purchase could be slightly different and/or use different wire colors. Please read your instructions carefully and just use this as a guide.)

That is basically it. Most shaved door handle kits come is multiple channel configurations. The spare channels can turn on lights, pop trunks and do any more operations you want on remote control to show off the vehicle. The limits are endless if you have the skill and imagination to use the full power of the module.

Wire Protection

You will be running wires through the bodywork to the doors for windows, locks and opening systems. These wires must carry a high current and will be quite thick. You must be careful to allow enough length to open the door and have them well covered with some kind of boot to protect from chafing and pinching in the door frame. For this we highly recommend a wire loop kit. This loop is basically a stainless steel spring with a 3/8" to 7/16" inside diameter. The spring is tough and flexible

and with one end fixed in place on the vehicle door, the other end can slide in and out of the car body when the door is opened and closed. The wires are then well protected and will not get kinked in the door frame.

Basic Power Door & Window Circuit

OK. Now let us look at the circuit. Let's start with the wiring on a basic system—one that allows the driver to control all four windows on the car and can lock out the controls on the other three individual windows. Each window will have its own motor and a switch for the person sitting next to it. In between the passenger switch and the driver's control switch is a switch that locks out all the passenger windows except to the driver.

In the diagram on page 77, the power is fed to the driver's door through a 20-amp fuse. You may choose to use a circuit breaker, a popular option. Power windows are using a lot of power and if they get stuck, the power draw can be more than the fuse can handle. While protection is essential, losing a fuse is an annoyance. But a circuit breaker will heat up and open the circuit then close it when it cools, and the window will operate again. Less hassle and just as good a protection device.

The power comes into the window-switch control panel and is fed to a contact in the center of each of the window switches. Two contacts, one on either side of the power contact, are connected to the vehicle ground and to the motor. The power also runs through the lockout switch to a similar window switch on each of the other doors.

When the driver presses one of the switches, one

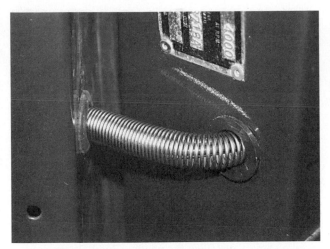

From this angle you can see how the spring protects the wires and looks very trick. The hard plastic allows the spring to slide easily.

You can take accessory power for these power door systems from the fuse panel or you can test for a wire that has power only when the ignition switch is on run. We used the fuse panel because it is handy and simple.

If you want to tap into another wire for power, get out your multimeter and check for power while the ignition switch is in the run position. If you find one, turn the key to off and see if it is still hot. If not, use it, but if it is hot both ways, forget it.

of the two side contacts is disconnected from the ground and connected to the center power contact, while the other one remains grounded. This provides power to the window motor. If the switch is pressed the other way, then current runs through the motor in the opposite direction and it changes the direction it is moving.

Power Locks—Here are some of the ways that you can unlock car doors that have central locking:

• by pressing the unlock button inside the car
• by pulling the knob on the inside of the door
• with the keyless remote entry system

A cable connects the power-door-lock actuator (solenoid) to a latch. It is very simple and effective. There is also a manual latch opening safety system inside the car to allow opening the doors when there is no electricity.

The power-door-lock actuator is a pretty straightforward solenoid device. It only pulls in one direction and when released is spring-loaded to release the pressure holding the door latch mechanism open. In custom and hot rod vehicles that use power door locks, the lock/unlock switch actually sends power to the actuators that unlock the doors. But, on late-model factory vehicles with more complicated systems, there are several ways to lock and unlock the doors and a body controller decides when to do the unlocking. A controller module takes care of a lot of the important little things that make a vehicle friendlier.

For instance, it makes sure the interior lights stay on until you start the car, and it beeps at you if you leave your headlights on or leave the keys in the

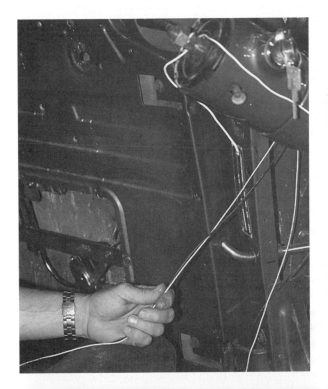

Once you have it located, run the wires to both sides of the vehicle for the door power wires, as well into the center of the cab for the switches in the soon-to-be crafted console.

This is what each door should look like. One wire to the door solenoid then another three for the window motor. It is possible to piggyback the solenoid and window grounds taking only one into the cabin.

There should be a pile of wires in the center of the cabin after you run the wires for the door locks and windows. This pile includes door locks, power windows, radio and speakers plus a hot lead for everything in the area.

On the '56 Ford truck the window lifts had to be loosely installed before installing the window itself. Be careful and slide the glass into the two pinch clamps that hold it tight to the lift.

Once the glass is tight into both clamps, remount the mounting strap that holds the lift system to the inner door panel.

ignition. For the power door locks, the body controller monitors all of the possible sources of an "unlock" or "lock" signal. It monitors a door-mounted touchpad and unlocks the doors when the correct code is entered. It monitors a radio frequency and unlocks the doors when it receives the correct digital code from the radio transmitter in your key fob, and also monitors the switches inside the car. When it receives a signal from any of these sources, it provides power to the actuator that unlocks or locks the doors, but only if the brakes are on and/or it is in park. This type of system is on many OEM components.

Multiple Door Systems—On really modern cars now there are so many electrical functions that power windows work in a completely different way. Instead of them being separate with power for the motor going through the switches directly, the switches are integrated into an electronic module in the car of which the average car contains maybe 25. Some cars have one in the driver's door, as well as a central module called the body controller. Cars that have lots of controls on the door are more likely to have a setup like this. They will have the power window, power mirror, power lock and even power seat controls all on the door. This would be too many wires to try to run out of the door. Instead of trying to do that, the driver's door module monitors all of the switches. For instance, if the driver presses his window switch, the door module closes a relay that provides power to the window motor. If the driver presses the switch to adjust the passenger-side mirror, the driver's door module sends a packet of data onto the communication buss of the car. This packet tells the body controller to energize one of the power-mirror motors.

On the more complex systems, not even the dealer's mechanic will fix the system in-house.

Instead they will replace parts and send the defective parts back to the manufacturer who can send them out to be repaired and then sell them again.

If you own a late-model vehicle, you can troubleshoot the basics of these systems yourself, but once you get past the basics, forget it. See a specialist.

If you are a hot rodder or enthusiast, there are many kits available, some more complex than others. It is quite possible to do your own installation and enjoy the special feeling these luxury features give you in your hot rod, street rod or street machine.

AUXILIARY DRIVING LIGHTS

The term driving lights will mean different things to each of our readers. Off-road guys will think of 1,000,000,000 candlepower lamps on the roof of their Baja rig, while others will think of additional fog lights for a better look at the road. Regardless of

The last step is to adjust the window lift tracks and the window so it slides freely up and down. This may take a few tries, but it isn't difficult.

Installing an extra set of driving lights is not as simple as it would seem—you must comply with federal vehicle lighting guidelines. Most auxiliary lights are to be installed so they only work when the low beams are on, and not on with high beams. These on the pickup truck are actually off-road lights that are never turned on unless the vehicle is off the road.

their power or how they are mounted, the following is a clear guide that will take you from the box they come in to being able to switch them on inside the vehicle. And, although this set of lights might be small and is being mounted on the front of an S-10 pickup, the methods are exactly the same as those used on stock vehicles, hot rods or Baja runners thundering through the night across the desert.

The starting point is to examine the mounting system that came with your light kit. It will probably be a simple mount that attaches the light with screws or bolts. Before drilling or mounting, make sure that the lights will be perfectly aligned by keeping the mounts at the same height and that they are evenly spaced from the center point of the

vehicle. On the Chevy S-10, for example, there is a bit of a point on the nose with the grille in the middle running from the center to each side. This design dictated that the mounting points be determined by placing a long straightedge across the front of the truck. This was done with a spare piece of 1 x 6 that was 6 feet long. The line it gave us dictated the mounting points for both lights. We scribed a small line on the top of the bumper where the lights would be mounted.

Follow along as we go through a step-by-step installation of a set of off-road lights beginning on the next page. These instructions (except for the wiring) are basically the same for any auxiliary light system.

INSTALLING AUXILIARY LIGHTS

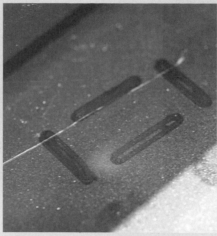

1. After placing the mount on the scribed line, we used the mounting guides inside the mount itself to mark the locations using a felt tip pen. Then we were able to choose where to drill the two holes required. The bumper on the S-10 is very hard steel so we used a high-speed drill with titanium-tipped bits. Once the holes were drilled, we placed the mount back over the guide marks and screwed in the two sheet-metal screws supplied.

2. Here you can see how we kept the mount aligned and parallel to the scribed line. Doing it this way allowed us to make sure the mounts would be pointing straight ahead.

3. Make sure you place the two sheet-metal screws to the inside edge of the mounting slot. By doing it that way if you need to focus the light more to the right of the road, the mount will be able to adjust.

4. The bulbs only have two wires coming out of the back. White is power and black is ground.

5. We fitted the bulbs into the mounts turning the mounting bolts just snug. That would allow us to adjust the lights later.

6. After we did both sides it was time to do the wiring that will connect the lamp power wires and send it to the light relay also supplied. Since these lights draw a lot of power, a relay is used to switch them on from inside using a small low-amperage switch.

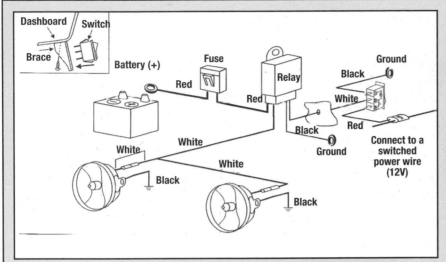

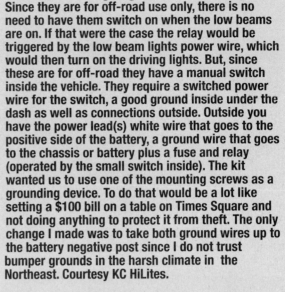

7. This is the wiring diagram sent with the lights. Since they are for off-road use only, there is no need to have them switch on when the low beams are on. If that were the case the relay would be triggered by the low beam lights power wire, which would then turn on the driving lights. But, since these are for off-road they have a manual switch inside the vehicle. They require a switched power wire for the switch, a good ground inside under the dash as well as connections outside. Outside you have the power lead(s) white wire that goes to the positive side of the battery, a ground wire that goes to the chassis or battery plus a fuse and relay (operated by the small switch inside). The kit wanted us to use one of the mounting screws as a grounding device. To do that would be a lot like setting a $100 bill on a table on Times Square and not doing anything to protect it from theft. The only change I made was to take both ground wires up to the battery negative post since I do not trust bumper grounds in the harsh climate in the Northeast. Courtesy KC HiLites.

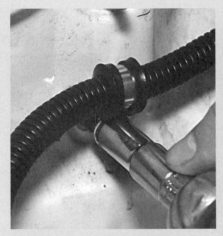

8. Both the power wires and ground wires for the lights were protected by plastic split tubing from right under the bumper to the relay that we mounted to the inner fender near the brake booster. By doing it this way we eliminated any risks from rocks or other debris.

9. The relay only requires a single screw for safe mounting. Keep it simple and the system will work better. You now need to run the two remaining wires (small black and small white) inside the cab so you can connect to an on/off switch. Carefully drill through the firewall making sure you do not hit any wires or harness, A/C tubing, etc.

10. In this photo you can see the wire being run to a fuse panel and connected with a spade connector at the accessory position. You can tap into the main fuse panel and run a wire from the panel to the switch. The wire right above the "ACC" is where we plugged into a switched power source. To make sure it was switched we tested it with the switch off and then on.

continued on next page

Installing Auxiliary Lights, continued

11. Once the wire is connected in the accessory position, you can install the on/off switch in the dash somewhere close to the driver. The black wire goes to one side of the switch (usually the bottom side or right side when it is horizontal) and the other side goes to ground. The 12v power wire goes to the middle. There should be enough black wire to make both connections. The ground is what allows the relay to become active and do its job. It is the relay that turns on the driving lights, not the switch. You can see that the switch is integrated into the dash and looks like it came from the factory.

12. From this angle you can see that the switch is in the bottom of the dash area where the headlight switch is located. I wired the switch so it will only come on if the headlights are on. This prevents me from turning it on accidentally when exiting or entering the vehicle.

13. Now is the time to see if both lights turn on. If they turn on and off properly, you need to wait until dark and then park in front of your garage door and turn them on for aiming. Make sure they are both aimed at the right side of the road on street vehicles. Also make sure they are aimed down, just below where the regular lights are aimed. (If it is an off-road vehicle like a Baja Runner, aim them straight out in front of the vehicle and get the maximum lighting possible to help the driver avoid obstacles and serious problems. This is how mine are aimed.)

INSTALLING AN AUXILIARY GROUND-WIRE KIT

Auxiliary ground-wire kits became more popular with the rise of the sport compact car movement. All of these vehicles are unibody designs and that type of construction has inherently poor grounds. Then, when hot rodders started adding 500 to 900 hp engine modifications, high-technology engine management systems and mega-watt sound systems, the electrical systems began to fail, partly due to overload and poor grounding. Many owners found it easier to add a supplemental auxiliary ground-wire system where they made duplicate and improved grounds that were installed into circuits without removing existing grounds. Today there are about a dozen companies selling these ground wire kits.

VE-Labs is one of the companies that makes auxiliary ground kits. Although all vehicles could benefit from an auxiliary ground-wire kit, they are especially useful for unibody, computer-controlled vehicles. Mike Green, developer of this product, has designed it so the radiator and heater core grounds go through a diode to make sure current gets out of the components to ground, but doesn't get back in. (This is a patented feature.)

This simple step-by-step installation process of the VE-Labs kit took about an hour on a late-model FWD sedan. It should take about the same for any vehicle, regardless of size. Just remember that a great deal of your vehicle's performance depends upon good grounds just as the safety of your aluminum components do.

1. With the master junction block mounted to the vehicle in a location that is convenient and visible, you can start running the ground wires to the locations where they are needed. Keeping them in sight will allow you to verify that they are not damaged and it will also keep them out of the weather for the most part. Two screws make it a simple installation.

2. The kit comes with wire separators that allow the simple installation to look very professional.

3. Each wire has a braided wire covering to avoid abrasion damage, an insulated layer to stop conductivity and the core wires. Before you start to install one, cut it to the proper length to reach the connection desired, slide on the shrink tubing and pull back the braided outer layer, and strip enough of the core wire to make a good soldered or crimp connector connection. You can choose to install the shrink tubing over the crimp connector or just up to it.

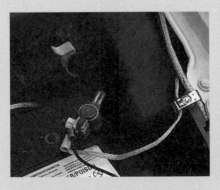

4. This cable was run to the chassis and you can see the shrink tubing that goes tight up to the crimp connector. This is a good chassis connection point.

5. I could have run a cable to the battery, but I ran the flat-braided wire to it instead. It didn't matter because they are all grounds.

6. This overview shows how easy the kit is to install. One wire to the chassis (frame and body or two places on the body if it is a unibody), one to the battery negative cable, one to the heater core, one to the radiator, one to the computer ground, one to the exhaust near the O$_2$ sensor and the last to the engine.

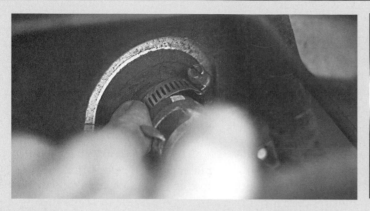

7. Installing the ground wires to the heater core can be difficult as it is all inside the vehicle and just the inlet and outlet pipes are visible. The trick to making the ground stick to the pipe is to use a hose clamp and strip the wire insulation and stainless covering leaving the internal wire exposed about one inch.

8. In this photo we are installing the ground wire on another pipe just to show you how it looks. There is no need to ground the AC pipes so we removed it after the picture taking was completed.

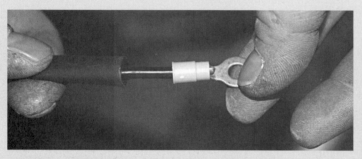

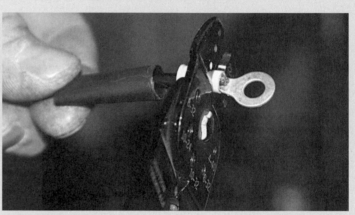

9. To make a ground wire use a terminal with shrink tubing, starting by slipping the tubing on the wire before you crimp the terminal. Note the three items: shrink tubing (left), wire already stripped (center), and the terminal in place but not crimped. This is a 10-gauge wire that went from the body of a custom pickup truck to the chassis. There are two of them—cab to chassis. There is nothing wrong with having a few too many grounds.

10. Next crimp the terminal to make the installation permanent. Note the shrink tubing is already in place. This low-cost crimping tool is OK, but if it had been done in my workshop I would have done a better job using my ratchet crimper.

11. Now is the time to heat the shrink tubing and watch it shrink and cling to the wire and terminal. We are using a special tool from Snap-on that uses butane to heat the tubing, but you can also use a match or an electric heat gun.

12. Although the shrink-tubing size selection might have looked too large, the magic of heat and this special plastic allow it to shrink three sizes smaller making it a perfect fit. On grounds it may not be necessary to worry about shorts, but you do need to worry about corrosion and damage. The completed wire is professional looking, and will last a long time.

Chapter 8

Installing Premade Wiring Harness Kits

One of the projects we'll cover in this chapter is a nice example of an original 1971 Plymouth Barracuda. The present owner had it more than 10 years and has fought electrical gremlins from the date of purchase.

Making your own wiring harness isn't as much a necessity as it was in the '60s through the '80s. Back then, there were few suppliers offering generic wiring harnesses kits for just about any application. Today there are many companies that will not only send the harnesses, but also a prewired fuse panel, pigtails for headlights and other connectors, outstanding instructions and all the fuses, crimp connectors and grommets you should require for your own installation. Because of that, there is little reason to try and make your own wiring harness for a complete car or truck unless you absolutely must. On motorcycles, instrument panels and a couple of other circuits it makes sense, but not for any big jobs. It simply will cost you far more than the kits cost to build your own and it will take you ten times as long. Even experienced car crafters will start with a generic kit because they know the value of time and resources.

REWIRING A BARRACUDA MUSCLE CAR

This segment is on replacing every wire in a classic muscle car. That's right, every single wire in the car will be pulled out and all new wiring harnesses will be installed. If you've got the car stripped down to the rails, there's no better time to replace the wiring.

From an electrical standpoint, anyone buying or owning an original muscle car from the 1950s to the mid '70s should automatically change out every bit of wiring in the vehicle. That may sound drastic, but it will solve many problems, improve the value and make the car safer to drive. It is

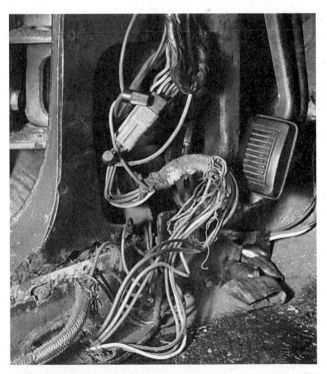

Most unrestored classics, or cars more than 20 years old, will have original wiring that will look similar to this—which was the case with this 'Cuda.

Before you begin any major how-to project, you need to get your tools organized. Ideally, you need to have a toolbox that will keep things in easy reach and that can be rolled easily. Keeping tools in order and easy to find makes the job go much quicker. This Snap-on box has room for all the hand tools I own.

When we removed the old fuse panel this is what we found. Taped splices and loose crimp connectors. This is not what you want on your classic muscle car. It is these kinds of things that have been done that make changing every wire in the car an imperative project.

important to remember that these cars are all more than 30 years old and wiring, just like everything else, breaks down as time goes by.

Of course the original and previous owners may have also made poor repairs and bad modifications to the electrical system also, just like on this 'Cuda. Today, almost every muscle car from this era has some specialty company selling new OEM accurate replacement wiring harnesses for the entire car. They are available for Chevrolets (and most GM muscle cars), Fords, MoPars, AMC and many others. The bottom line: there is no reason to fight electrical gremlins when new wiring is just a phone call or click away.

The project for this chapter is an original 1971 Plymouth Barracuda. The present owner, Mike Priore, has had it more than 10 years and has fought electrical gremlins from the date of purchase. Mike calls it a driver, but it is 100% OEM and too nice to be a real driver. He had done his homework by looking up all the option codes from the build plates attached to the left front fender, as well as making sure the VIN number was correct. This 'Cuda has once been modified for drag racing, so it was missing a lot of the original options. Mike's goal was to restore the vehicle as closely as possible to original condition.

Although this is a very nice car, it gets driven a lot, and the wiring is either original, meaning old,

or it was replaced some time ago, and not by a professional. Mike was continually having electrical problems when we approached him for the project. As a Honda mechanic and long-time enthusiast, he was up for the task himself. He had already looked at some of the wiring and found potential shorts to the body that could have caused a fire. He originally thought of making his own harnesses, but decided it was easier to go with new OEM-quality replacement harnesses that were readily available. It is much more cost- and time-effective to purchase a new harness rather than make your own. If you would like to make your own, or if one isn't available for your vehicle, see Chapter 9 on how to custom make your own harness.

We called YearOne and they supplied every harness available for the 1971 'Cuda, as well as some of the missing parts, such as driving lights. This included the master fuse box/bulkhead assembly pre-wired for installation with the master harness under the dash. Each unit sent by YearOne equaled the OEM product in quality and accuracy.

Getting Started

The first thing to do is to get good wiring diagrams. Mike had purchased the "Jim Osborne Diagram" book for this model a few years earlier and it was a lifesaver. It is important to note that while most of the remove-and-replace process is intuitive, there can be a few difficult bumps in the road to completing the installation. Be patient and think everything through before acting (don't cut any wires until you know exactly what you are going to replace them with). A typical example was the transmission-mounted reverse light switch and indicator light on the dash. We could not get them working when we changed the harness. That switch turns on the backup lights. Making the connection was easy, but getting it to work took some thought. We discovered that it wasn't the switch, it was the reverse gear rod adjustment.

The dome lights were not obvious either. If the car hadn't been butchered by the drag racer, it would have been an easy unplug-and-plug deal. However, the wiring didn't exist and we had to spend time on the wiring diagram to find that the wires actually go to the rear of the car and then back up.

The rest of this section will be a step-by-step photo how-to detailing the removal and installation of every single wire in the vehicle. Although we are demonstrating on a particular muscle car, the procedures and techniques are generic enough to be applied to any early muscle car, classic car, street rod or hot rod.

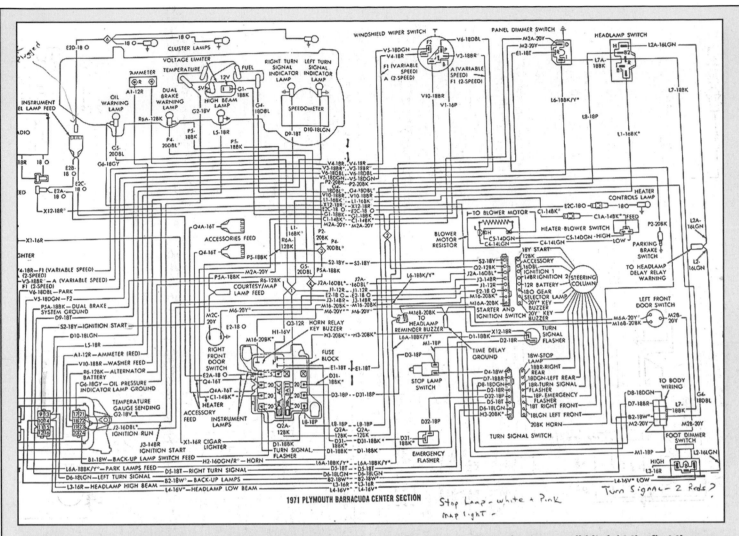

1971 PLYMOUTH BARRACUDA CENTER SECTION

1. This '71 Cuda wiring diagram manual by Jim Osborne is what we used for every step to make sure we did it right the first time. Having to do things over was not an option. Working without good wiring diagrams can make the job impossible if previous owners had modified or improperly repaired the existing wiring. These and similar manuals are available on the web.

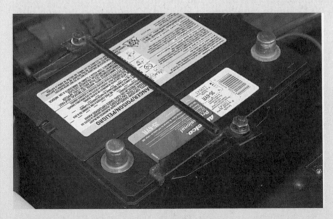

2. Disconnecting and removing the battery first should be obvious. Make sure you take off both the positive and negative cables. Do not attempt to change wiring harnesses with the battery connected. Make sure you remove both cables completely, because circuits can find a ground through the body if they are bare and the positive cable is connected. Removing just the ground cable is not enough.

3. Everything in the interior of the vehicle needs to be removed. This includes seats, carpeting, all instrument trim panels and all dash panels. Tag, bag and label the fasteners in relation to where they go and what they connect to. Get the seats, console, carpeting, shifter handle and any trim pieces off the dash around the instrument cluster.

4. Do not take the dash instrument cluster out yet, just remove any face plates accessible from the outside of the dash without going under it. You can remove, bag and mark the fasteners from the heater control panel, radio, speaker covers, etc., but do not take these items out of the dash yet.

5. Remove everything from the rear of the car also. This includes the seats, carpeting, trunk carpeting, hard panels mounted between the rear seats and the trunk, spare tire, floor hard panel spare cover, jack and wrench, and anything else that could make finding and removing wires, bulbs and other things that might be connected to this harness difficult. (Rear defroster for example.) The spare should be placed out of the way with the seats.

6. When removing the carpet you will remove the door sill plates. (They are aluminum coverings that hold the carpet in position at the doors on either side.) Once the driver's door sill plate is removed you will find the rear harnesses running front to back under it. The 'Cuda had a metal plate over them to make sure they didn't get damaged. We found three cables going back to the rear. Two original and one home made. (It is the one with the electrical tape cover.) Behind the cables you will see the steel cover that protects them underneath the sill plate.

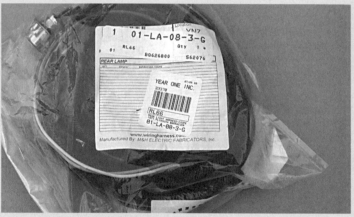

7. Each part from YearOne is tagged with information that tells you what it is. This is critical so you get the right part installed in the right place. Pay attention to this information.

8. We decided to do the rear harness first because it is fairly easy to find and get to. Additionally, since it is easy, you will get used to the plug-and-play installation technique. The first step is to follow the harness up to the front of the vehicle and identify where the big connector connects to the master fuse panel or another connector up front. That is where the rear harness plugs into the system.

9. As we followed the rear harness back through the rear seat area, there was a short harness cable coming from the kick panel on the driver's side next to it. It is probably for the overhead dome light, the missing rear defroster or something else, but it is not part of the rear harness.

10. This is the damage-prone area that the factory didn't engineer well. Running the harness over the support that ties the floor into the body is poor engineering. It leaves the harness prone to damage by rubbing/abrasion and/or impact from kicking as passengers get in and out of the rear seats. It needs a hard protector to stop potential damage—we made one up from old harness split tubing. Once again this car had two OEM harnesses going to the back plus another that was made up. (Probably the electric fuel pump near the tank for drag racing.)

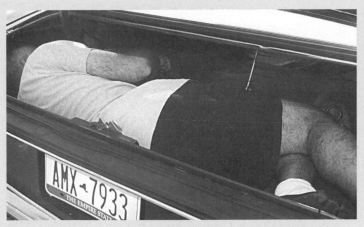

11. After looking at the task, we determined that we would run the rear harness from back to front as the first step. You will need to get someone in the inside of the car helping as you run the new harness around body supports or other obstacles and get it from the trunk to the front connector at the kick panel. Remember, you are installing it from back to front, just laying it in place right now.

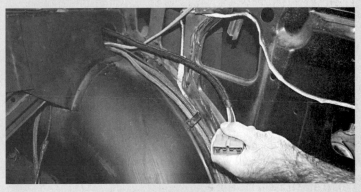

12. You can push it through and they can pull it into the interior area. It is far easier this way than it would have been trying to feed all the small wires and connectors through the small hole between the interior cabin and the trunk one at a time. Back to front is the key. This is where the harness goes up and over the rear wheel well. There are clips along the way to keep it in place. The new harness came through the bulkhead with a bit of work and gentle pulling. Once the big connector came through it was simple. Keep in mind that we have not touched the OEM harness yet.

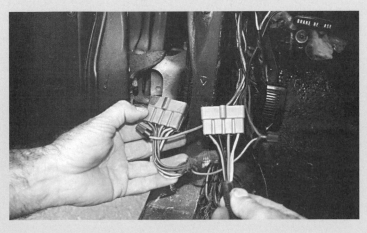

13. Once the main connector was inside the car, we pulled it up to the front and connected it where we disconnected the old OEM harness. Remove the old rear harness from the front connector and plug the new harness right into that front harness connection. (Even though you will be changing the front master harness later, go ahead and plug it in now.) The rest of the installation of this harness is remove-and-replace one item at a time, going from front to back. Once the rear harness was disconnected from the front master harness, we started working our way back to the rear trunk area. Inside the passenger compartment was simple, we just had to reroute the new harness to where the old one was. Here we are bending the clips up and removing the old and setting the new one in place.

14. Any place there is another harness next to something plugged into the old harness, just unplug the sub-harness from the old one and plug it into the new one. (You will come back and run the new sub-harnesses after taking care of the rear harness first.) Once you have it connected in the front inside the car, you are ready to spread it across the entire trunk and start identifying the taillights, brake lights, license plate lights and directional signal lights one at a time. The rear harness goes from the driver's side to the passenger side by routing the remaining harness and connectors across the rear of the car underneath the trunk lid at the bottom of the entrance. Remove the old harness from the clips and insert the new harness.

15. In the trunk we removed one light at a time and replaced it with the new harness and a new bulb. Just go from one side to the other and it is a simple deal. Start by removing the side marker light on the passenger side finding its match on the new harness. With a new bulb installed, insert it into the left side marker light housing. Most bulbs can only fit in one way, so look at the socket and the old bulb holder and install it right the first time.

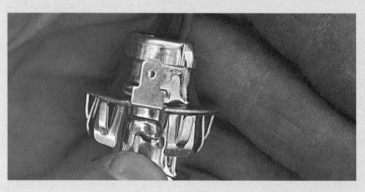

17. Now it the time to replace every bulb in the car as you remove and replace the wiring harnesses.

16. Some bulb connectors have a slot or other type of guiding system so be patient and take it one at a time. The part of the harness that comes with grommets indicates that it goes outside the trunk itself. (The fuel tank sender connection is one example.) Pay special attention to these, but you may want to do the other connections first, since you are already working in the trunk area.

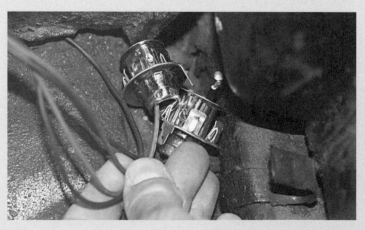

18. Be very careful as you remove lights from housings in the trunk. You will find that there may be several housings that are assemblies from the factory, not just bulb holders. These will connect to part of the new harness going back outside the trunk. (We used a small 12 volt cycle battery and small wire jumpers to test these housings before we re-installed them.)

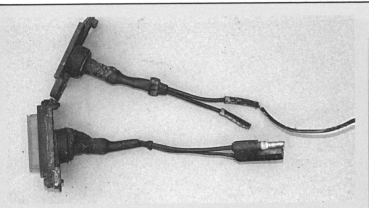

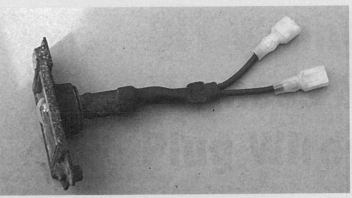

19. (Left) We found one marker housing assembly (driver's side) had the factory connector removed and replaced with crimp connectors. To repair this we had to make up waterproof spade connectors. (Right) If these housings are in poor condition, replace them with new ones. (Also available in NOS or OEM reproduction replacements.) After fixing everything else, including the housings, install the wires with grommets going outside the trunk by pushing the wires through the trunk and gently pushing in the grommets. The new grommet should be lightly lubricated with a non-petroleum lubricant (liquid soap is great) and pushed into place, through the floor. Insert the bulb and bulb holder into the empty housing or in the case of the fuel tank, make that connection and go to the next.

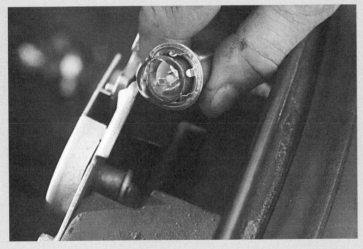

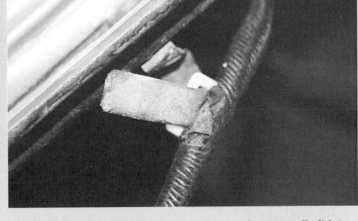

20. The process is simply unplug-plug-and-play. As long as you do it wire-by-wire and bulb-by-bulb, it is foolproof. Just take your time and keep going.

21. After all connections are made to the new harness, clip it into place as you remove the old one. Follow the new harness back toward the inside of the car and you will see a slotted white piece of plastic inserted into the harness. That slides over the clip at the top of the trunk compartment just under the driver's side of the trunk lid. Slide it into the clip and fold the entire clip below the edge of the compartment. Then you can pull the old harness out from the trunk area (guiding the big connector until it is in the trunk).

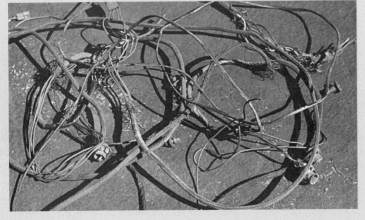

22. This is the mess we replaced in just the back of the car. The old rear harness has corroded bulb sockets that didn't work, shorted wires to ground that luckily didn't burn anything but the fuses. There were repairs made in several places and overall it smelled nasty and was nasty. (That's a good enough reason to replace it, not even counting the electrical problems.)

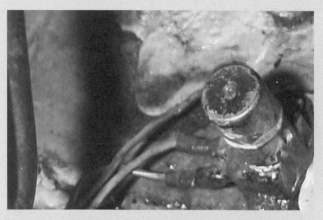

23. A foot-dimmer switch is prone to damage, which is the main reason why OEM's moved it to the steering column. Even though it is not on the bottom of the floor, a lot of moisture from the driver's feet will get to it. The accumulated years of dirt make it very ugly.

24. This is the original foot-dimmer switch after 35 years and it was really nasty. It was replaced at the same time the harnesses were changed. YearOne has these in stock for most cars. Note the bent connecting pin, which was the cause for the faulty operation.

25. The next step we made was to get the front harness out of the package and lay it out on the driver's fender and across the front of the car past the hood. It was clear that the front harness includes every light in front of the vehicle. Once again we routed the new harness from the front of the car to the bulkhead connector so we would not have to pull little wires through one at a time. We started at the passenger outside headlamp and pulled the harness through to the left headlamps and then to the starter solenoid and finally to the bulkhead connector.

26. We did the front light and accessory harness next and then we installed the engine harness. In both cases we ran the harness from the components in the front of the car to the main bulkhead connector and went forward from the front to the firewall, removing the old connectors and replacing them with new ones, one at a time. The installation was fast and easy. In our case, because of missing options on this ex-drag racer, we had several connectors that didn't connect to anything. This shot shows the front harness and the four connectors you will have on each side. Starting at the top and working down: master headlamp (three-prong connector), high beam headlamp, side marker light (yellow) and park/directional light (black).

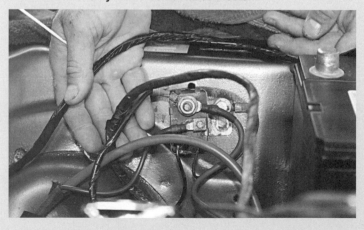

27. Here you can see the old front harness still connected to the starter solenoid. The fused wire that should be attached to the top bolt is missing and has been replaced with a non-fused replacement whose crimp connector is broken.

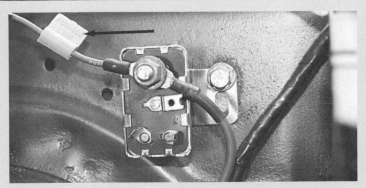

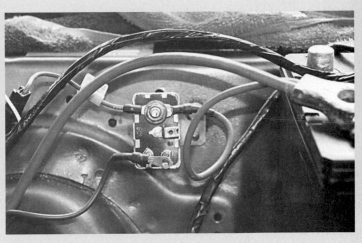

28. With the new harness and solenoid installed, you can see the OEM fused wire as it should be (arrow). It is a factory "fusible link" that protects the wires inside the car if there is an overload from a bad starter, bad starter gear or anything else that could create a huge current draw condition. This is important, so never fix a burned-out fusible link with plain wire; you would be placing your vehicle in great danger as it can burn every wire in the car in an overload condition.

29. Here is the completed solenoid installation. The only wire we need to connect yet is a factory ground cable which we will leave for last.

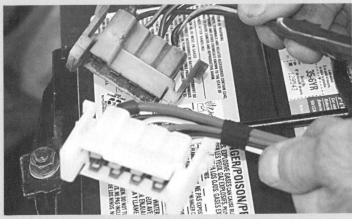

30. Next is to install the new front harness connector to the bulkhead connector.

31. As you are installing the harnesses, make sure all grounds are clean of paint and rust. Here you can see a before photo that shows a ground that was in really poor condition. Any component depending upon this ground wouldn't function well or possibly not at all.

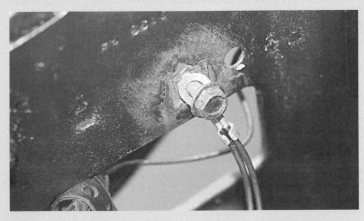

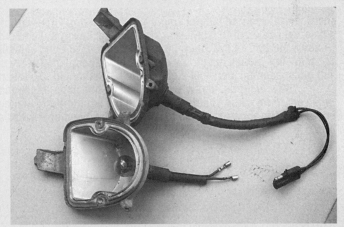

32. Some sandpaper and elbow grease repaired this ground surface. You can use touch-up paint to improve the appearance as we did when we completed the installation.

33. Now, getting back to the reason we asked you not to throw away the old rear harness yet. When we pulled out the front park/directional lights we found two different connectors on them. The one at the top is the OEM connector and the bottom one is something a previous owner had devised.

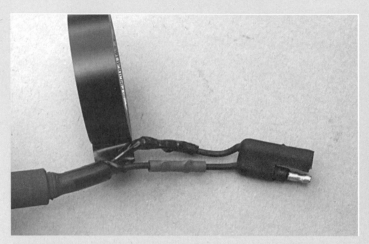

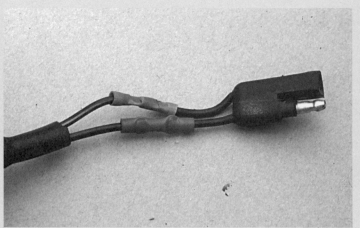

34. We took the correct connector off the old rear harness and spliced it into the wiring of the parking/turn signal light assembly. We taped it to make sure it would not be able to pull apart. Notice that we taped one wire by itself and then taped the entire wire group back to the light assembly. There was no way to use shrink tubing here.

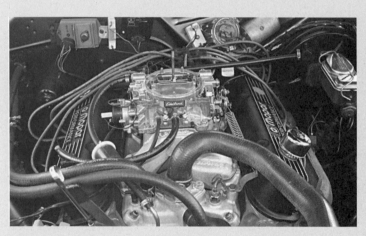

35. The ignition from the factory was electronic and the distributor is without points. There was also an OEM high-performance Chrysler ignition box that Mike had installed a few years back. Plugging it into the bulkhead connector wasn't needed because at this point we needed to unplug everything from the bulkhead connection box so it could be removed with the master fuse panel inside.

36. We followed the same plug-and-play procedures for the engine harness. The old one had all the correct connectors, even though they had been spliced into for one reason or another. It went in easily and after we had it in, we disconnected the new engine and front harnesses as well as the two small plugs that control things like turn signals, wiper motor, etc. With these all disconnected from the bulkhead connector, we were ready to go under the dash.

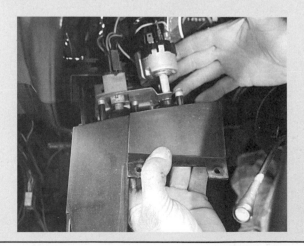

37. The first part of the dash wiring we touched was the windshield wiper switch that is mounted on an easily removed panel. Removing this and unplugging it allowed us to pull the connectors from the new underdash master harness to where it goes in easily.

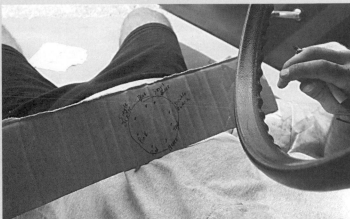

38. The next step was to remove the instrument cluster enough so we could see behind it in order to make a wiring diagram of the connections. Since the actual cluster was for a Rally Dash and the master wiring diagram we were using was only for a standard dash, the new diagram Mike made was critical. This way, we were able to make our own diagram and return the car to this configuration as we installed the harness.

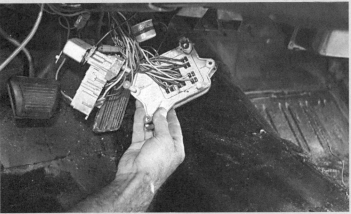

39. Once we had the instrument cluster diagram made we removed the cluster and removed the bulkhead/master connection box before we installed the new one. In the right photo you can see the non-stock wiring coming out of the back of it. We are holding the new one next to the old.

40. This photo is of the hole where the fuse and bulkhead connections reside. They go between the engine compartment and the inside of the car. That is the throttle cable to the right of the hole coming back towards the lens.

41. The bulkhead/master fuse connection box, once removed from the bundle of wires of the harness, had burned wires, indicating an overload and a few non-OEM connections to the box. A perfect scenario for a fire. We cut out the bulkhead/master connection box before we made all the connections because with both in place there would have been far too much clutter under the dash.

42. The bulkhead connection part of the assembly was fine, but we changed it anyway.

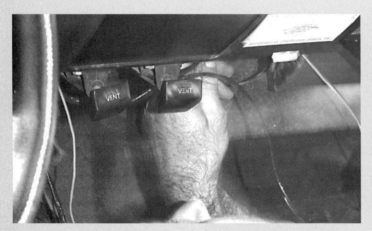

43. Once we installed the new bulkhead/master connection box, it was a slow, but easy process to take each connector from the old harness and replace it with the new harness. The wires were then pulled out of the vehicle. In many cases there would be several connections on the same wires in the harness. We did the "unplug/plug and play" game one at a time and got it right the first time.

44. There were several connectors that we couldn't install because of the missing accessories. That included the speaker wires, radio connector and a silver box that we found on top of the heater. After researching it, we found it was part of the stereo system. We removed it for now.

45. The job was completed from an electrical standpoint once the underdash harness and bulkhead/master connection box were installed, followed by the underhood connectors to the bulkhead connector. Now, after you have the instrument cluster installed and the windshield washer panel connected, you are ready to test the system. After the battery is connected, start testing each circuit; windshield wipers, horn(s), radio, heater/AC fans and clutch, electric windows and locks. In other words, test every circuit including the ignition and starter. Do it one step at a time and have someone standing by the battery to remove the positive cable if needed. It should start now.

WIRING A PRO STREET TRUCK

Although some would argue that the Pro Street era is over, there are still a lot of enthusiasts building them. From pickups to Novas, skinny tires in front with monster street slicks in the rear are still burning power from mega-cube engines. Part of the attraction to building these vehicles is the "anything goes" philosophy that allows the builder to be as creative as they wish.

The order in which we wired the Ford pickup in the following photos might seem a bit unorthodox, but waiting for parts, paint, windows, bedwood, and other details dictated what we could do and when. Most home car building projects work out the same way.

The first thing is to decide how many circuits you will be using and then get the appropriate premade generic wiring kit. We selected a Painless Performance Wiring's 18908 18-circuit, GM steering column Universal Street Rod Kit, which includes instruction manual #90501. This manual is really helpful and easy to follow. You may not need to do everything shown in it, but you MUST read it chapter-by-chapter all the way through. We cannot show you every single step in the wiring of a car and any drawings for your wiring harness and instruments would not be valuable to you either. Every hot rod project has its own set of challenges.

Since we were doing the job using Painless Performance wiring we also used their accessories exclusively. Please look at the photos to see what we used on the project.
- Remote Electrical Shutoff
- Chevrolet Light Switch
- Battery Remote Location Kit
- Gauge Wiring Harness w/Electric Speedometer
- Neutral Safety/Backup Switch
- Fan Relay w/Thermostat Kit
- Hot Shot Kit
- 4 Switch Dash Panel Kit
- Manual Electrical Shutoff
- Accessories/Brake Switch, Alternator Pig Tail and PowerBraid Harness Protectors.

1. This beautiful 1956 Ford was built over the course of a year by Steve Agnello and Tom Peliatire. They did the chassis welding, bodywork, paint, installations and interior modifications. We were lucky enough to be allowed to wire the truck starting with nothing but a chassis and driveline.

2. In true Pro Street fashion, this '56 Ford pickup includes a late-model Camaro front clip, 454 cid big-block Chevy, aluminum heads, MSD Ignition, COMP Cams flat-tappet hydraulic cam, Edelbrock Victor manifold, Holley 760cfm four-barrel, a TH-400 automatic trans and many other parts from other vehicles and makes.

3. The engine had already been broken in (20 minutes at 2500 rpm to break in the camshaft, flat tappet lifters and bearings) and was ready to start up and power the truck. All that had been done at this point was decide which wiring kit we were going to install and the rough layout of the harnesses we would create.

4. From an electrical point of view, working with a body-off frame and chassis is ideal. We were able to carefully design and lay out the electrical system in the best possible locations without interference from body panels. The hardest part of the project was waiting for things to get done so we could do the wiring.

5. The Painless Performance instruction manual is organized into a systems approach, and shows you how to do the front wiring, rear wiring and underdash wiring step-by-step. With a sedan, only the front wiring goes outside the body. In the case of the pickup truck, the wiring for the rear of the truck needs to go outside and under the body to the rear where the directional taillights, brake and license lights would all be mounted. The fuel gauge wiring also had to go outside of the body, but it was stopped in the middle so it could be attached to the sending unit from the fuel cell to be mounted in the bed. The truck will also have the battery mounted in the bed and a ground that goes directly to the chassis while the positive wire goes to the starter and a junction block up front for distribution to the rest of the vehicle.

6. It seemed like the kit came with 10 miles of wiring coiled up into a fuse panel/block assembly. There were also lamp connectors, pigtails for the alternator and a bunch of extra fuses along with a bag of crimp connectors and electric ties. Every wire has a number and name indicating where it goes. They are all color-coded, but the names are easier to work with. But, if you are adding wires at a later date it would be wise to match the colors of the kit. (Save all cut-off wiring from the kit during installation as you will find it useful later in the installation and since it already has the correct name on it, you can save problems later in the life of the vehicle.)

7. It can be very intimidating at first until you start separating the wires by where they go: front, rear and inside under the dash. Then you can start separating them by what system they connect to, outside as well as inside. (There are several bundles already tied together from Painless Performance and it is best to leave them connected until you are ready to run them.) Your main concern is to install the harnesses as cleanly and neatly as possible. It is imperative that you take your time and measure everything carefully.

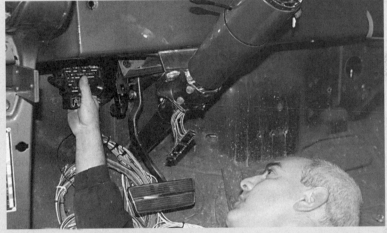

8. The next step is to determine where you want the fuse panel mounted. About 90% of the time it will go on the left side of the firewall and you will drill the hole for the wires to exit the body on the firewall near the master cylinder. But you can mount it anywhere you want as the wires are plenty long to even mount it on the right side.

9. Get under the dash and fit the fuse panel. In our case we loosely installed the brake pedal assembly and the emergency brake assembly as they had to be in to make sure we chose the correct location, and placed the firewall reinforcement on the firewall where we knew the fuse panel would fit. We marked and drilled the holes from the inside out using a small drill bit. Note: Too much pressure on the drill will only slow things down and you will end up doing damage if you are not careful.

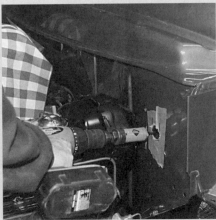

10. Although we started by drilling the mounting holes from the inside out, you will want to drill the mounting bolt holes to size and the big hole from the outside in. This is really a "measure it three times and cut once" type of deal. We covered the area with masking tape before starting to drill with the hole cutter. It results in less paint damage and a very clean hole for the grommet. Use the firewall reinforcement plate as your pattern for drilling the big hole (before you return it to the inside and mount it under the fuse panel). Line it up with the holes already drilled for mounting the fuse panel and then mark the big hole.

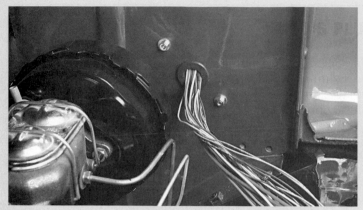

11. Once that is open, you install the grommet and start sliding the groups of wires that go outside the cabin through the grommet. In our case it included all front and rear wiring, engine wiring, fan wiring and fuel gage sender wiring as outside wires. All of this was stuffed through the grommet before bolting the fuse panel in with the mounting bolts from the inside.

12. At this point we had all the outside wires sticking through the grommet, the fuse panel installed and we started separating front from back. Once again do it two or three times, checking each wire by name to know where it is going. We draped the outside wires over the inner fender and went to work. Remember, these will be configured into a harness going back and forward and having to do them over again is a lot of hard work and frustration to go back and fix them.

13. Since only the left inner fender was installed and the right inner fender, fenders and front end sheet metal were not installed, we did the rear harness first. Most of the work is under the truck, so jack it up and use solid safety stands before getting under it. We had the left inner fender on before starting the wiring because we needed to support the wires going down to the underside of the truck as well as across the front of the truck. We began configuring each wire harness, double checking each wire name and rechecking where it was to go.

14. Here you can see the wires we split into forward and rear wires, as well as see the wires still remaining inside the truck. The wires inside will stay there, while the others will become a front and rear harness. You can see the front fenders on a fixture for sanding and reworking them.

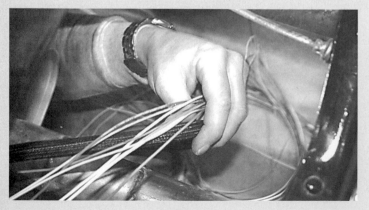

15. We ran all the rear wires down the inner fender to the underside of the truck first. Then we wrapped them all with the Power Braid and pushed it up from the bottom of the truck into the grommet on the firewall. We then ran the protected harness back towards the rear by mounting it to the top of the inner-left frame rail, where it would be protected from most dangers on the road. The rear of the cab was the first place we could get the harness up where we could see it.

16. At that point the wires came out from the rear of the body. These are the wires before we covered them with the Painless PowerBraid protection covers. We started the PowerBraid from the back of the cab and ran it to the rear of the frame inside the boxed sections. The pink wire is the fuel sender wire; remember it for later.

17. We separated the fuel sender wire and bundled the rest using tape to keep everything together. We then connected the bundle of rear wires to an electrician's wire puller so we could capture it and run it through the first section of boxed frame rail on the way to the rear.

18. There is a short section (about 12") of unboxed frame required to deal with mounting bolts in the frame for suspension components. These wires are now past the first section and ready to be pulled through to the rear. To complete the pull we drilled a 2" hole in the back of the frame and pulled them through after installing the PowerBraid. The completed wiring harness is invisible from the top of the frame and well protected.

19. That was all we could do in the rear until the bed was mounted.

20. While the front fenders and sheet metal were being installed, we wired up the headlight switch and the steering column-controlled circuits. The first step for the headlight switch was to determine the best location. This truck had the original switch on the left part of the dash, but the GM switch we ordered with the kit didn't fit there. We considered other locations, but in the end decided to modify the switch to fit rather than drill another hole in the dash.

21. To trial-fit the headlight switch, we pulled out the on lever and tried to mount it on the dash as it was. Removing this pull lever requires you to push down very hard on the spring lock as you pull it out. Put the switch in from the back and then install the locking collar from the front of the dash. In our case it wouldn't tighten down because of the recess in the dash.

22. To get it to fit and look good, our do-everything technical guy, Pete McGregor, machined a small spacer that fit into the recess in the dash which allowed the switch to fit flush. (We could have cured the problem with two washers of the correct size, but machined aluminum is prettier.) The headlight switch was now ready to install. But we also had to wire the connector that is included for this job.

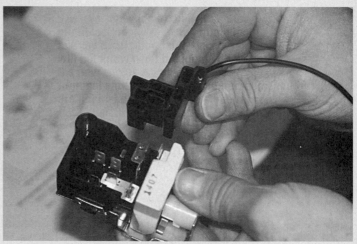

23. The connector slides right on the switch, so connect the wires with the terminals attached that say headlight switch to the connector. We connected the ground wire before we did anything else.

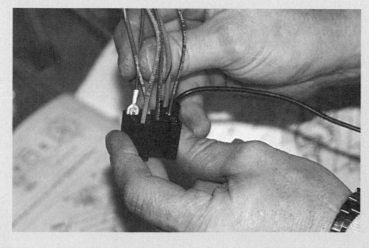

24. The final step after connecting all the wires into the light switch is to snap the connector on, then mount it in the dash. You can now forget about this part of wiring the lights.

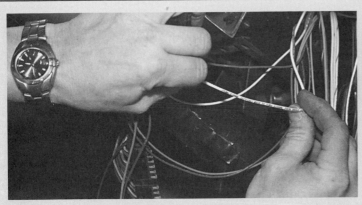

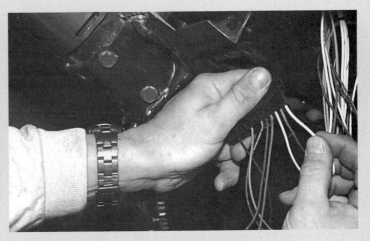

25. While we were still inside working, we wired up the steering column connector. If you ordered the correct parts as recommended by Painless' tech support, this is a very easy task. In the box you will find the correct connector for the steering column you are using. In our case it was a 1974 Camaro column (the same donor car that supplied the complete front clip). The instruction book calls out the wire codes/color for each wire and you simply stick them in the right pin positions to match the column. Every wire for this connector will refer to the column in some way. For example, the turn signals say Turn Signal Section.

26. Hold the connector in your hand and snap in the pre-wired terminals into their respective slots. Once you are done, snap the two connectors (male and female) together. You can see the color match of the wires you connected. Most of the time OEMs are the same color that Painless has used, but read the instructions to be sure. In general, GM column connectors follow this color coding: Brake Light (white); Left-Rear Brake & Turn Light (Yellow), Right-Rear Brake & Turn Light (dark green), Turn Signal Flasher (purple), Emergency Flasher (brown), Right-Front Turn Signal (dark blue), Left-Front Turn Signal (light blue), Horn Ground Wire (black) and Shift Indicator Light (gray, column shift only).

27. The wires are already terminated with the matching connectors to the column connector

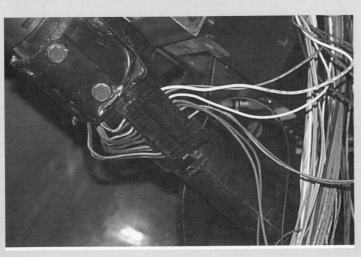

28. Next snap the assembly into place on the steering column. When they are together in place you will see how the colors match almost perfectly with the OEM wiring. You will also note the three wires that are in the OEM connector and not the one from Painless. Those three wires are for the windshield wipers and washer. The '56 Ford windshield wipers are vacuum manual but electric wipers without the washers will be installed later.

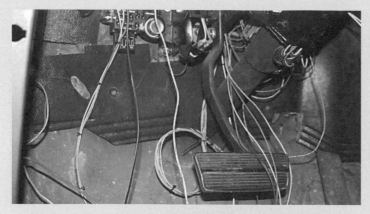

29. While we were under the dash we slipped the brake switch into the hole on the brake pedal assembly. Rough-adjust it so the pedal just has to move a little to close the circuit—turning on the brake lights. As the pedal is pressed, the spring-loaded switch pushes the contacts closed. When you release the pedal it opens the circuit.

30. Knowing which wires are the correct ones is simple: look for the brake light switch wires and follow the instructions. We used insulated spade terminals which actually were a bit of overkill, because the switch is unlikely to get wet, but it is better to add more protection now than have to trace or fix a problem here later.

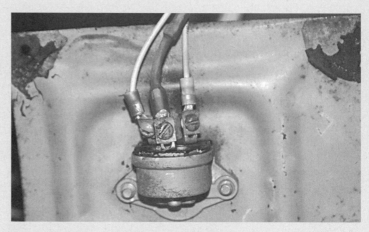

31. We only added one wire to an accessory position on the fuse panel during the entire installation. It was for the radio and was the smart place to get power for it. Yes, this is before we finalized the installation by wire-tying all circuit wires together and then placing the bundles up out of the way.

32. We also rewired the stock headlight foot-powered dimmer switch and tied it into the correct wires under the dash. The owners wanted to keep it instead of using a column-mounted unit. Again, the car won't be a daily driver.

33. The bodywork has now been installed so we can move forward with the rest of the wiring. We needed to install the wires for the ignition, starter, electric fan and thermo switch, plus a relay to power it all plus the alternator, electrical master current quick disconnect, headlights, parking and turn signals with hazard flasher, and package it into good-looking harnesses along the way.

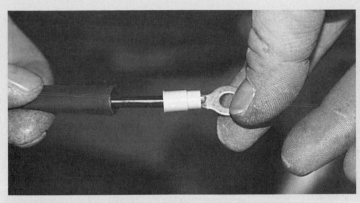

34. But before we did that we built ten ground straps, three of which mounted between the body and the chassis/frame. (We installed a kit on page 87, but we'll review part of it here.) They were made up of black #10 wire with eye connectors on each end. We made them all 12 inches long and to make them look good we also shrink-wrapped the ends. Start with 12 inches of #10 black wire and strip each end to the depth required by the crimp connector.

35. Then slip on the shrink tubing and then the yellow eyelet connector. (The colors are for different size wires so get it right the first time.) You are now ready to crimp the connector. The crimping tool can be like this low-cost unit we had at Steve's garage or I could have brought my ratcheting crimper. Just make sure you crimp using the right size on the tool—yellow eyelets to yellow crimp location, etc.

36. I use a simple propane torch from Snap-on to shrink tubing, but you can use any kind of heat source, even a 1500-watt hair dryer can work in a pinch.

37. The completed crimp connector connection is solid and safe and will last a very long time. It didn't need to be made waterproof, but shrink tubing makes for a professional, neat appearance and strengthens the connector.

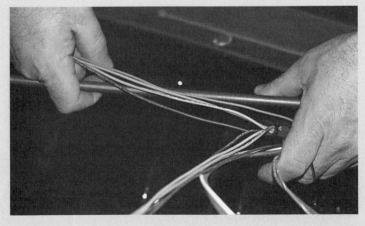

38. To get started with the rest of the wiring, we once again reviewed every remaining wire by name and where it would be going before we actually did any wire routing or cutting.

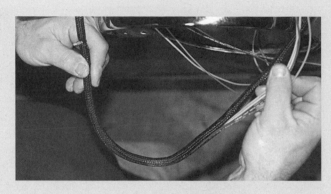

39. Once we had the systems separated, we made our plans for the front harness. The front lights (high and low beams, direction signals, hazard lights, parking), fan, horn, alternator, alternator exciter wire were all bundled together. We also included the A/C and oil pressure and engine-temperature sender wires even though we were not using them. (No A/C is planned, and the oil pressure and water temperature are direct-reading mechanical gauges without senders.) If a new owner decides to use senders, or install A/C, the wires are right there in the harness bundle and the wiring notes indicate them.

40. The other harness was the engine harness and it includes the distributor, tachometer, starter solenoid, fan temperature thermal switch and battery feed. They were bundled and run across the firewall, below the back of the intake manifold with the ignition to the distributor and fan-temperature thermal switch wires coming out of the harness in the middle and running to their termination points protected by smaller PowerBraid.

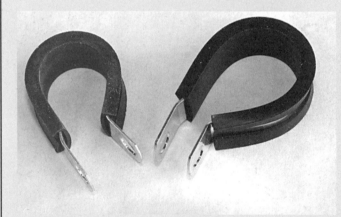

41. The harnesses were attached to the lower seam weld by Adel insulated clamps to protect the wires and keep them from rubbing anything. We used more than two dozen of these clamps in sizes ranging from 3/8" to 3/4". They are even used to hold electric inline fuel pumps to the chassis, to fuel filters for the same reason, and your creativity is the only limit to their applications.

42. In this photo you can see the thermal switch for the electric fan on the bottom middle (insulated female spade connectors) and the wires in Power Wrap going up to the HEI-type distributor. The keyed starter wires went directly to the starter solenoid.

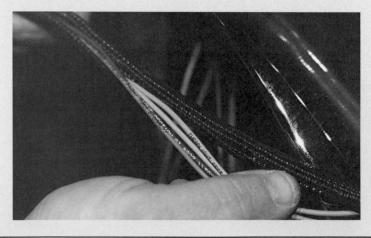

43. The front harness was then run up to the radiator support as a unit and split into right and left wires at that point.

44. The left wires went into the headlight and parking light housings while the right wires went to the right side of the truck with the fan wires and the alternator wires coming out of the harness at their appropriate locations in the center of the radiator support behind the grille.

45. Once the headlight bulb holders were in, the bulb connectors were spliced on and the parking, turn signal and flasher lights were also connected. These are all actually the same light; they just operate with three circuits and two filaments in the bulb.

46. These before-and-after shots show the instant change in the look of the truck once the front end was wired and all the lights were installed. There were no loose wires showing, just nice compact and beautiful black harnesses. The light trim rings had not been installed and it still looked good to me. However, we still weren't finished yet.

47. Changing to direct-reading gauges for water temperature and oil pressure required some additional lines. For the oil pressure, we had to run an oil line from the side of the block, bottom left. We made a braided hose cut to the proper length with the proper-size connections. The water gauge was more of a problem, as it is hard-wired together from the factory, so a bigger hole needed to be cut into the firewall.

48. The answer was another grommet made by Steve. It was actually a body plug from something that had a hole drilled in it for the cable and was then slit to get the cable into the middle. It turned out very nice. I'm sure someone makes these things already modified, but when you get a group of old-school hot rodders together, this is the type of innovation that you'll see.

49. We now had to finish up a bunch of small wiring details for the front of the car. The first was the installation of a horn relay in the front of the car and a circuit breaker for the fan itself. On the left you can see the horn relay we added to the circuit between the power wire and the horns. After we completed the circuit we found that the horn wouldn't honk and it was because there was already a relay in the circuit and the one we added opened the circuit. So we left it there and ran the thermal switch for the fan to it and it turned on the fan automatically for us. Every time the circuit is closed at the switch the relay lights off the fan.

50. Next to the circuit breaker we mounted the 75-ampere master fuse and holder. Power comes from the rear-mounted battery to the starter and then to this fuse before it goes to the rest of the circuits needing power. There is also a master junction block close to the starter, as the starter gets power directly from the battery without any fuse in the circuit. But the starter solenoid power comes from the fused circuit at the master fuse panel. (These details are spelled out in the instruction manual.)

51. This wiring harness that started at the master fuse panel and has come all the way across the front of the truck under the radiator support and to the power junction and alternator. It might seem strange to run it across the front instead of under the car, but we wanted it very protected as it powers just about everything in the car. Likewise, we wanted the fan power to come from close to the master junction block, after passing through the circuit breaker, on the right of the car and going to the middle of the front end. The harnesses are beautiful and functioned the first time.

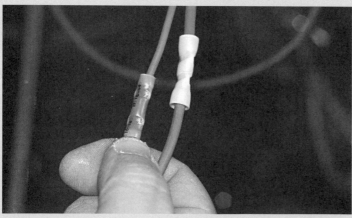

52. Anytime you splice, solder or crimp a connection, don't be afraid give it a test tug. Better it break now than when the car is assembled or driving in freeway traffic.

53. The alternator needs to be wired before we finish with the engine compartment. We got out the pigtail purchased for this three-wire alternator installation. (One-wire installations are even easier.)

54. The connector snapped together with the wires and we snapped it on the alternator. Then we ran the red 14-gauge wire to the small power terminal on the back of the unit. The white wire was sent to the power red 10-gauge wire post. These small wires are important because an alternator needs 12 volts to excite the fields of the windings in order to make it create AC power. Once created, the diode triad converts it to the 12 DC the vehicle needs.

55. This is the original heater box, fan motor and cables from the '56 Ford. It didn't look too bad at first, but things deteriorated once we opened it up.

56. Pete removed the fan motor and it was really in bad shape. The guys in the shop wanted to reuse it, so they rebuilt it by cleaning out all the rust, soaking the bushings in 30-wt. oil and actually got it to spin again when power was applied. We did rewire it with fresh wires and plenty of shrink tubing. (The three wires are for Hi/Low and Power.)

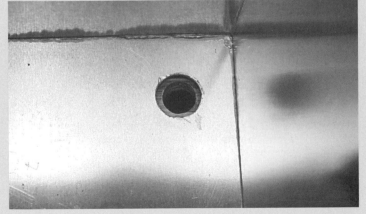

57. Now it is time to install the battery in the back of the bed. An aluminum battery box was purchased and it was bolted solidly to the hard oak flooring of the bed. Holes were first drilled through the aluminum and the oak flooring and then the aluminum was ground away with a rotary air burr so the only thing touching the battery cables is the wood and wood is an insulator.

58. Pete improvised by using a cutting torch to sweat-solder the terminal end for the battery ground cable. A bit of overkill for the job at hand, you would need a lot of skill to avoid melting the wire. A propane torch would have been better.

59. Although Pete did anneal the copper a bit, it was just fine for a battery cable that was going to the frame 12 inches away. A little bit of shrink tubing made it look great. We found him a propane torch to do the positive cable, however.

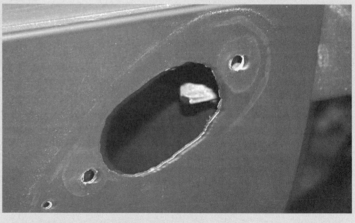

60. The last steps are to clean the paint off the area where we were mounting the cable with a rotary burr and then bolt it in with a 7/16" bolt, lock washer and nut. While we were under the frame, we added a few of our extra ground straps between the frame, roll pan and bed.

61. With the roll pan grounded we were ready to install the '39 Ford taillight reproductions. These two wire assemblies take a double filament bulb and are for brake lights, turn signals and emergency flashers. The hole for the lights was cut through the roll pan before it was painted. Any less than perfect hole-cutting was covered by the light assembly, thank goodness.

62. The reproduction taillight assemblies are better than stock. Note that the studs on the light base are the grounds for the assembly, as the white and black are not grounds.

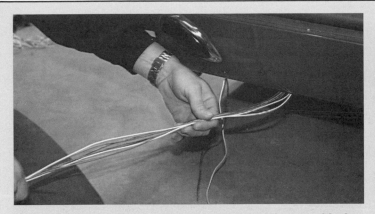

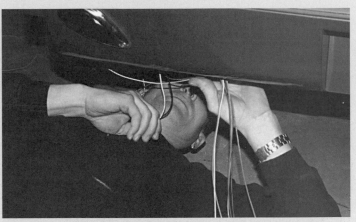

63. Pete is holding six wires in his hands: two yellow, two black and two green. They are all that is needed to complete the wiring in the rear of the truck. Just sort them out by length as well as name. (The longer go on the right side.) One of each is for the taillights, one is for the turn signals and one is for the license plate lights. Refer to the instruction manual anytime you get confused, lost or simply want to double check things.

64. It just took a few minutes to wire up both sides in the rear.

65. At this point we went back inside the truck to complete all wiring, before the team started installing the interior. This included the dash part of the instrument cluster, some loose wires we needed to chase down and the doors, windows and radio/speakers. Pete began with the wires that would go into the 9-pin connector that allows the dash insert (instrument cluster) to be removed without taking out the gauges. This took about 50 minutes because the terminals were the very small ones that really take time or the correct crimper.

66. It is a good idea to test circuits as they are installed to make sure they are working before you start burying them under panels and carpet.

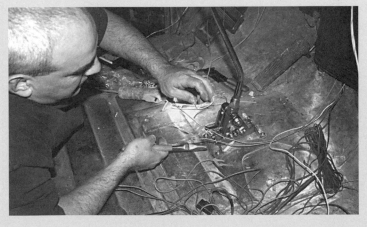

67. Steve is running the wires for the doors, windows, radio and speakers before he starts construction of a center console. He also has to wire up the park/neutral safety switch. The console and carpet will hide all of this so only we will remember how ugly the floor is.

68. This is the instrument panel, finally installed. We thought this subject deserved its own chapter. The install begins on page 154.

69. It is hard to believe that the entire wiring system for this truck came in just a few boxes. It is a very satisfying feeling to complete the job and hear the engine start up and watch the gauges come to life.

The electrical systems of race cars are obviously quite different. Rather than hide all wiring like you would in a street car, racing electrics need to be accessible for quick repairs.

WIRING A RACE CAR

Wiring a race car is a lot different than wiring a street car. Although many parts of the job are the same, the overall objectives are what spell the difference. For example, on a street car you want all wires, fuse boxes and computers out of sight. On a race car you want them visible for quick access when something goes wrong. On a race car you hard wire everything and even run some redundant systems that can be switched either by the driver or automatically if the primary system fails during a race. On race cars, fuses and circuit breakers must be designed to be changed quickly during pit stops or between runs. And, all systems need to be strengthened or protected so they don't fail during the hard-driving conditions. Race cars produce a lot of vibration and heat, and the occasional impact.

So, before you start ripping out all the factory wiring, check your sanctioning body rulebook to make sure it is okay to modify it. Some allow for complete wiring mods, some are only partial—it will depend on the class and sanctioning body.

It is obviously easier to start with a bare chassis, but in the case of a stock-class race car, where all or part of the OEM wiring must remain as is, the job of installing performance electronics can be quite challenging. That was the task we were faced with on the project racer in the following pages. Not only did we have to check every connector in the vehicle and replace anything that seemed marginal, we also eliminated most of the functions from the old OEM key system.

We removed every chassis ground wire and created an alternate grounding system that takes every ground directly to the negative terminal at the battery. This is to reduce the possibility of developing a bad ground. Intense racing can put a lot of stress on chassis grounds. Additionally, since we know every ground is going directly to the battery, it is much simpler to fix if a ground problem should develop.

At this point we will illustrate how we custom-wired this race car to conform to the racing sanctioning body rules.

1. We ordered a Painless Performance Race Kit, Street Legal, 12-circuits with an 8-switch panel and the Painless Performance Wiring kit that is usually used for Pro Street drag racing cars that must run all street systems. The kit upgrades a lot of the stock wiring with a heavier gauge, and it eliminates any wiring for any systems that won't be needed for racing.

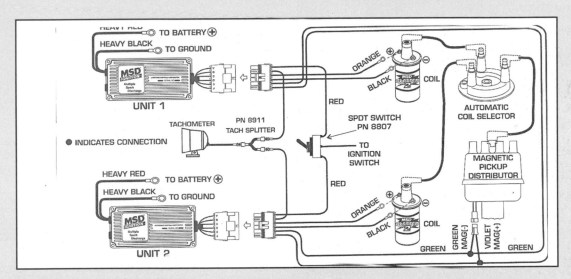

2. For example, every car running in one of NASCAR's big series has at least two ignition systems (usually an MSD ignition box/spark manager/coil) and if the ignition fails, the system will switch to the back-up system in less than a second. Running two ignition systems on a race car is commonplace and the switch-over can be automatic or with a manual switch like in this schematic. Redundancy on a race car is critically important and the additional weight is less important than reliability. Courtesy MSD.

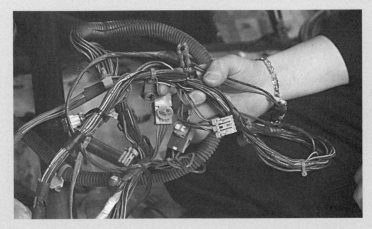

3. Racing rules allowed us to eliminate the back-up lights, license plate lights, interior lights, some of the stock gauges, and all of the unneeded harnesses and flexible conduit installed by the factory. We rewired all of the light wiring (brake and taillights in the back and headlights in the front). We also left the side marker lights and added wiring circuits so external hood-mounted racing lights can be installed for 12- to 24-hour endurance races. The handful of wires Louis is holding is just the first step in removing unneeded wiring and we had only gotten started at this point. We ended up reducing weight as much as 15 lbs.

4. We found a cracked ground connector on the negative terminal that was surely part of the cause of intermittent performance problems the owner had experienced. In the process of wiring this race car we removed and replaced every bad connector, every weak wire, and added grommets where wires went through the body into and out of the engine bay or to the wheel sensors for the antilock breaking system.

5. There were many problems found and each of them was repaired or changed altogether. These crimp connectors had burn marks on them and so did the insulation. This told us there was a short to ground in the old circuit.

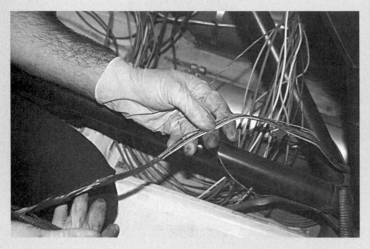

7. We started by disconnecting both battery cables and removing them completely. (Always remove both the positive and ground cables from the battery.) Not doing so can cause bare wires to short out.

6. Instead of fixing it, we replaced every wire in the back harness by making a new harness with the Painless Performance kit.

8. On ex-street cars that are now race cars you should remove the battery to see if any bulkhead openings are behind it. In the case of the Acura we found the main harness going from inside to the engine bay compartment. The original grommet was damaged and the wiring was starting to fray from rubbing against the firewall metal. In the back of this photo you can see the big black grommet. This is the place where all wiring exits and enters the cabin. Note that we had already replaced the bad ground cable connector for one that can handle several grounds to the battery at the same time. We wanted to move the battery to the rear of the chassis but the rules wouldn't allow it.

9. To the right of the battery is the main junction for power distribution. It has 50-amp fuses to protect circuits. It also has a 100-amp fuse for the starting system (the light-colored fuse on the bottom left). The battery buss and the main junction box would normally be on the left side in most American cars.

10. The engine ground was a strap running from a cam cover bolt which was insulated by rubber washers under the steel washer and a nut to a chassis ground. That is not going deliver a good grounded circuit. Having seen many performance problems, we knew the car would run far better with the new wiring and ground strategy.

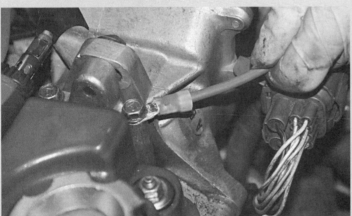

11. This is a good example of how we solved problems that needed to be corrected before we even started installing the new wiring and fuse panel. We disconnected this wire from the insulated stud and remounted it to the engine at a better location. We used the same ground strap for the revised engine ground, but we replaced the terminal ends and routed it from the head to the battery ground connector.

12. This is how the new battery ground cable looked after we added the engine ground strap. But, we also added more ground straps to it before we were done.

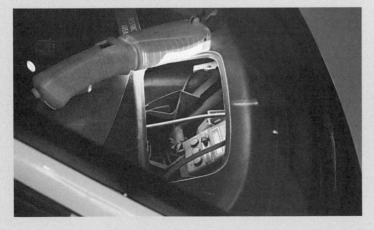

13. The first step in installing the new system was to mount the fuse panel and get the wiring from underneath the dash to somewhere it can be run to all required circuits. We were looking things over for a good location. I took a drop light (another required tool) and looked into an open hole on the dash. An airbag used to reside inside, but we found a strange electronic device. We asked the owner about it and it turned out to be the shift light control manager that allows each gearshift to be programmed for a specific rpm. This was perfect for us so we removed the shift light manager and brought it up out of the hole.

14. We knew that we wanted to get that device up on top of the dash, and we knew we needed to mount the master fuse panel on top also. The answer was simple, but required some fabrication. We decided to make an aluminum cover for the opening and mount both items on top. We started by using a piece of cardboard as a template. David traced it from underneath the dash while Louis held it flat from the top.

15. David added 3/8" all around the tracing to make it overlap the hole in the place that originally had a plastic cover and then cut it out. David then placed the template over the hole and it fitted the first try. He transferred the shape to a piece of scrap 1/4" aluminum and checked everything one more time before he cut it out. A little belt sander work to smooth out the edges and it was ready for the next step.

16. We then drilled holes for the fuse panel using the dash support template. From that we knew where to drill the hole in the back for the wiring. (Although the fuse panel is upside down in this photo it didn't matter when drilling the holes.)

17. To make sure none of the wires would rub against the aluminum plate we used a suspension bushing made out of polyurethane as a grommet. Besides looking very trick, it is the perfect grommet for this race application. The wires from the Painless kit fit through with ease and the other hole is for the cable going to the shift light sender and mounting holes for the fuse panel. The shift light sender is mounted with industrial hook and link using Velcro. Remember, we are hardening the electrical system to make sure it will always operate at peak performance. It is always better to stop problems before they start. The wire for the shift light manager goes through a rubber grommet for protection also.

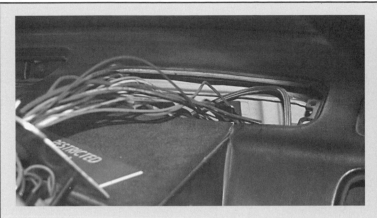

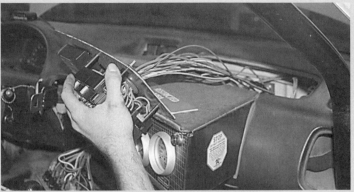

18. Here we are sliding the wires down into the hole making sure that everything is clean and safe. The braided hose in the back is for the direct reading oil pressure gauge. This simple modification has made diagnosing electrical problems during a race easy.

Although it is sitting without any restraint, the last thing done was to install a tab at each corner to keep it tightly in place during a race. This through-the-windshield shot demonstrates just how easy it is to identify blown fuses from outside the car.

19. This is how it looks installed. Clean, safe and easy to diagnose blown fuses, and isolate problems without having to get under the dash with a flashlight.

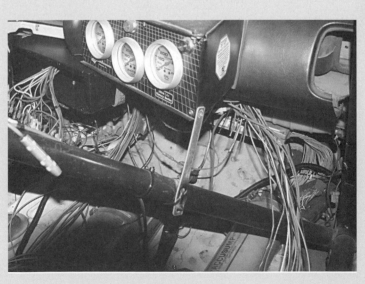

20. We ran all the wiring for the rear from the master fuse panel to the back of the hatchback. (You can see it at the bottom right of this photo.) It included taillights, brake lights, side marker lights, fuel pump and a few extra circuits for special things someone might want when building something from the ground up. We were also able to terminate a number of wires right underneath the master fuse panel. These included the extra circuits, rear window heater and back-up lights.

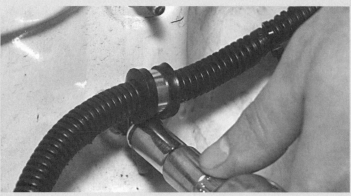

21. Here you can see the two taillight housings with the closest one already modified and the back one stock. The side marker lights and the directional signal lights were removed. You might also notice that these housings have 6 studs plus 6 nuts and lock washers holding them onto the body. We fitted them back in with just 3 and they are just as tight. An ounce here and there can make a big difference. Just don't remove fasteners unless you are sure they are not required.

22. All it took to modify the taillights was about 45 minutes (that was mostly running new wires and adding the protective split tubing). We took the new wires from the Painless Performance kit and ran them to the taillight housings, through split tubing to protect them as well as to keep it off the floor. There we spliced them into the correct wires (brake light and taillight) for each side.

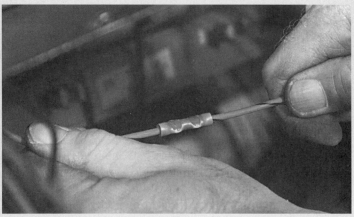

23. Once again, be sure you are using the right crimp connectors for the size of the wire. Notice that the Snap-on crimper also has them marked on the side of the crimping head to show you where to crimp them correctly. You can also just place the crimp connector into the crimping by matching the color of the connector to the appropriate color spot on the tool.

24. Once a crimp is made pull on it from both sides. Don't be afraid to pull hard, as it is better to have it fail now, not during a race.

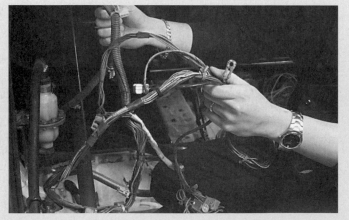

25. You will be amazed at the junk wiring you will find all over a race car that still has its original wiring. Weight is saved with the kit and potential trouble spots are eliminated. Here is how you harden a race car, electrically speaking. Once the lights were completed, we removed all of the old harness and connectors we were not going to use. The old wiring was not protected by very much split tubing, leaving it open to being hit and causing a short or an open circuit. Something like a forgotten wrench, loose equipment, people climbing over it, etc. The new wiring is all off the floor, in protective split tubing and safe from unguided flying objects.

26. We wrapped all wiring with electrical tape before inserting it into the protective split tubing. Some experts say tape can loosen as it ages, but inside protective split tubing it is safe, and outside it is easy to replace if needed.

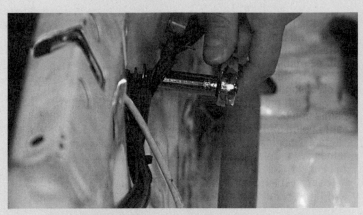

27. We removed these wires and replaced them with new. They were also up off the floor of the trunk using Adel clamps and split tubing that wasn't done stock. The wires were as safe as you can make them and still have access to them. Only a serious crash will damage them and then you simply have to start over again. It is a race car after all. We used Adel clamps on big wires without protective split tubing.

28. These are power wires to the system shut-off and ignition. Once again getting them up and out of the way is good insurance that doesn't cost much. (The Adel clamps cost about $1.00 each.)

29. It is important to properly terminate wires that you will not use. Here is an easy three-step process. We left enough wire in the terminated bundles so that at a later date if you need them they can be spliced into with ease. The first step is to get all the wires you want to terminate into a single group. By terminating wires you do not need right now, you can reduce the weight they represent yet retain enough of a pigtail to use them at a later date. You do this by cutting off the wire you do not need and leaving about a 12" to 24" pigtail in case you want to use it at a later date.

30. Next use diagonal wire cutters (Dykes) to cut each wire to a length just a bit shorter than the next. You may also do it with all the wires the same length if your cut is very clean and no wire ends can touch another wire. This precludes any possibility of shorts between wires. It is an easy procedure but it takes practice.

31. This photo shows Bert shrinking tubing on an underhood bundle of non-essential wires. He chooses to go from the back to the front, but both ends work the same.

32. In the Acura, access to the fuel tank pump and fuel sender are inside the car even though the tank is below it. These two items (small circle—fuel gauge sender at the top, large circle—fuel pump and lines) need constant attention in a race car. They both have access doors, which we have taken off for the wire repair work.

33. In this close-up of the fuel pump and hoses you can see that the connector is a simple 2-wire connection, power and ground. The small fuel hose is the return line and it is obvious that someone has worked on it at least once. (The hose clamp has been changed to an aftermarket style.) The pressure side of the pump (big hose) is still using the OEM connector. Both were fine and we wired the pump into the 8-switch panel we installed. In this manner the driver can both prime the pump before start-up as well as shut it off without shutting down any other systems.

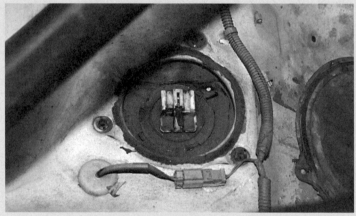

34. This shot of the fuel gauge sender shows us that it has three wires, one for ground, one back to the gauge and the third one to the low-fuel warning light. The connector was very dirty and required that we clean it out with a chemical cleaner and small wire brush. Although we were not allowed to change these wires to another gauge, we did repair the terminal connectors. We used the stock wires and male spade connectors to make the connections safe and strong. This included shrink tubing to support them right up to the spade connectors. Although I didn't install it, Joe agreed to have Bert install a pressure switch (aka Hobbs switches) into his fuel pump circuit so if the car gets in an accident it will shut off as soon as the engine quits running.

35. To have a simple and strong junction/ground post for all interior circuits, Bert welded this bolt to the chassis and we ran all of the interior grounds to it. In this manner after all interior grounds were on the post they probably had a good ground with the chassis, but we placed a 10-gauge wire on top and ran it to the ground post of the battery. This allowed us to meet our goals of not relying upon any chassis grounds to keep circuits operating after a crash or other type of incident.

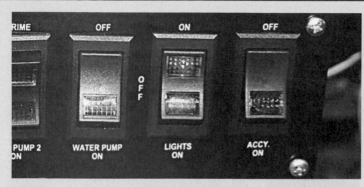

36. Getting the lights to work with the master switch panel we installed took a lot of work. There was a lot of bad wiring in these circuits that was hard to locate. So, we did it the simple way. We cut the wires off the headlight bulb connectors and spliced new Painless wiring into them. Although we couldn't remove all the old wires, replacing them in the circuit was very smart. They are larger, have never been smashed or damaged, and they were also clean.

37. Modern cars with the new covered lights use specific connectors to the bulbs and wires. They go in at strange angles so it is easier and much faster to leave a decent pigtail and cut the old wires off to splice in new ones. After crimp splicing each wire we covered that splice with shrink tubing. We pulled out as many of the old wires as possible and removed the hot leads so they couldn't cause problems. It was another way we could keep the weight down.

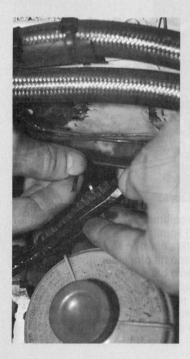

38. David uses wire ties on the new wiring onto an older one that was nearby. The taped wires between his fingers are for the add-on driving lights for night racing. That circuit was completed but not installed as the night driving lights were not available while we were doing this project. But, what we left will work with any system just by adding waterproof spade connectors and plugging them in.

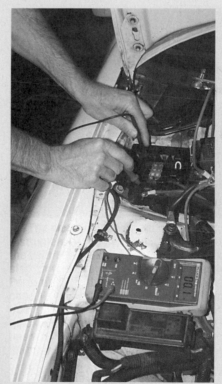

39. When we needed to check for power or continuity we used our trusty Fluke digital multimeter. It also gave us resistance with ohm checks as well as circuit integrity. Since we started by fixing all the circuits we were going to keep from the stock system, David spent a lot of time with the multimeter and adding new wiring instead of leaving the old mess. There were old wires everywhere and cleaning it up cured many problems even before we started installing the Painless kit. Never simply assume that a wire will have continuity, check it out as it just takes seconds once you are set up.

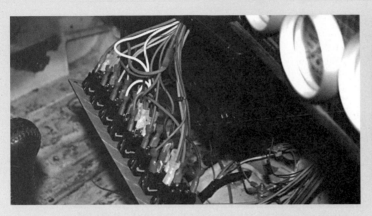

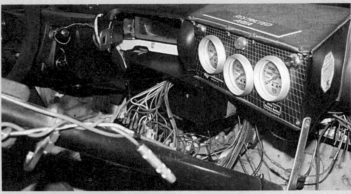

40. The last step in the project was installing an 8-switch master switch box which came prewired for easy installation. The hardest part was mounting the switch box, which eventually ended up just left of the instrument pod. It was installed with a Sawsall, 4 screws and one strap. This was kind of crude, but effective. You can see the wires going into the mounted black box for the switches into and under the dash.

41. We simply ran the wires out of the box and back under the dash where we could get them easily. All the installation took about 1 hour because we already knew where every wire was in the car, where to splice things and where to create new circuits. The lights took the most time, while the fuel pump took about 5 minutes. This is the how the switch panel looked when we were completed. You can see where the Sawsall did its thing. To be candid, it came out looking good, far better than I would have expected. The switch panel is within easy reach of the driver, which is a critical factor. So set the driver in his seat and mount it where he can reach the switches without having to stretch when you are fitting a driver control switch panel for yourself.

42. The last part was covering all wires and fuses with split protective tubing. On this car we used the hard type. Wire ties, split tubing and a lot of attention to the details resulted in a professional wiring job even though we had to integrate a complete kit with some of the OEM stock engine management wiring. Looking at the engine bay or the interior you will find all wires protected, off the floor (except for the antilock sensors and the wires for the fuel system) and looking like new. In the hatchback area it is the same story. The completed job improved everything about the car and even gave it a bit more power, thanks to the improved grounding system.

Chapter 9

DIY Wiring Harness & Plug Wires

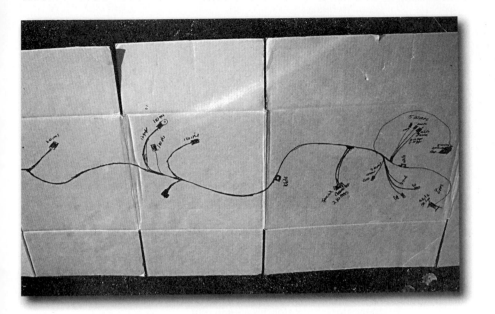

Although this cardboard box diagram looks a bit crude, it is the most important step to creating your own wiring harness. It doesn't matter if you are starting from scratch, or copying an old harness, the job is the same. You just start from different directions. Make sure every wire is to the size it will be when completed.

In the last chapter, we stated that it wasn't worth the time and effort to make your own wiring harness given the fact there were so many premade kits available. But, there may be times when you'll need to make your own harness; perhaps not for the whole car, but maybe for a subsystem, such as driving lights or some auxiliary audio/video system. Perhaps you'll have the odd car that doesn't have an aftermarket kit available, or you'll need to know how to make one for a race car. Finally, maybe you are the type who just likes the satisfaction of making things yourself.

In this chapter, we'll detail how to make an engine harness for a 1965 Dodge Dart with a small-block V-8 engine. The basic process is to take off the old harness and use it as a template for the new one. In another section, we'll also detail how to make a custom spark plug set.

Once all the drawing was done I started to strip off all the harness wraps so I could see below. This is a time-consuming process but worth every second. Doing this kind of a job requires patience when stripping away the covering and tape from the original harness so you can see beneath it and understand where each color wire starts and where it goes. Some of these wires are actually jumper wires from one connector to another, usually because of a common ground strategy.

As you keep removing the covering you may run into clip hangers as used by many OEM's to hold harnesses in a specific place, not able to move around. Once the harness wrap is removed and it falls off, put it right back into position with a tie wrap. Just keep taking off the wrap and use black wire ties to hold everything in position. You will find that you have to use several at junctions; don't worry about the wire ties, they are cheap.

Once all the harness wrap is removed and all the black wire ties are installed it is time to start changing out the wires that are bad and leaving the good ones in place. To do this we will replace the bad ones one at a time and when it is completed we will put on white wire ties to let us know these wires are fine.

You can copy a wiring harness as I did or you could start from scratch and create your own harness. In either case, I would stick to original wiring colors when copying and follow the Painless colors if using their 12-circuit fuse center (highly recommended if wiring a complete vehicle).

Their instruction guide that comes with the kit is outstanding. Matching their colors is actually quite easy, but if you have to use another wire color, just note it on the wiring diagram you are making as you add wires and connectors.

Starting from scratch, work from the component (starter solenoid, starter, oil pressure sender, etc.) to the fuse panel. By doing that you will be able to identify a route that will work as you add other wires and make a single harness for the engine circuits, rear section circuits, etc. Do the same for every component and those would include at least the following when creating a new main wiring harness for a vehicle from scratch. The Painless 12-circuit wiring center includes the following circuits:

• Alternator
• Battery
• Lights
• Starter
• Starter Solenoid
• Main Power Buss
• Heater
• Windshield Wipers/Washers
• Ignition—coil, distributor, ignition key
• Accessories—A/C, audio, video, games, windows, door locks, etc.

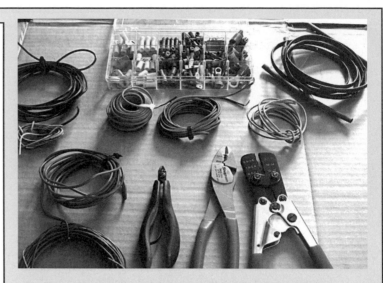

1. This is the cardboard and everything you should need to replicate the OEM harness. I am choosing to work with crimp connectors, but soldering would work just as well.

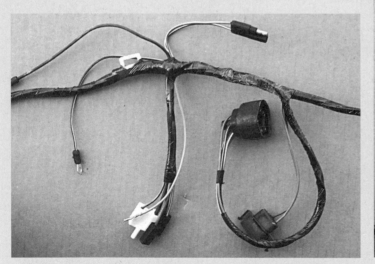

2. Once the harness is out of the vehicle the first step is on the large cardboard drawing board. Once it was open I took the old harness and laid it on the cardboard, tracing its shape and making notes about what kind of connector was in every position, including the number of wires, colors and a shadow drawing of what it looks likes. This information will make my job easier because I'll be reusing the connectors and all I have to do is match them with the right colors.

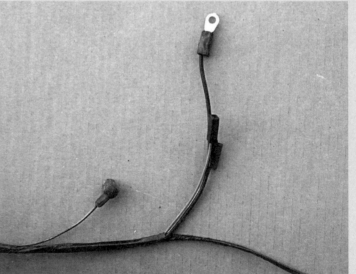

3. This part of the harness at the far end and is only four wires. Look at the next photo and see how I drew it out.

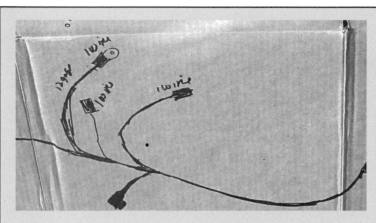

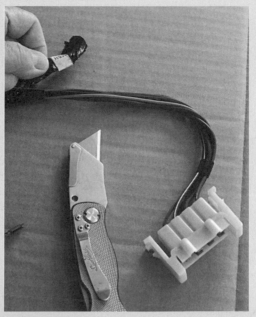

4. You can see how I gave myself the information. On the left there are three connectors. One is an "eye" connector that bolts on to the starter solenoid. The next two are simple one-wire slip-on spade connectors that are black and rectangular in shape. The last one on the left-bottom is a connector that slips over a threaded pin. I pointed the pin connector facing down as it comes from a different side of the harness.

5. It took about 30 minutes to carefully cut off the old harness wrap so I could see how things looked underneath. This is one step in the process you cannot rush.

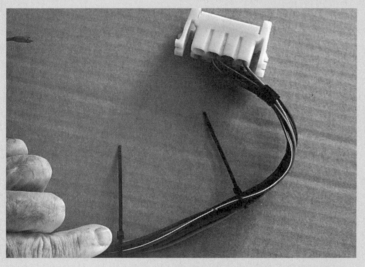

6. Place black wire ties every few inches to make sure nothing moves and you are simply undressing the harness, not changing it. (You will see why I said black wire ties in a bit.)

7. Although I agreed to make a new engine harness for my friend, the one I removed looked fairly new. Most of the wires were still flexible and the connectors looked good. Once I had the wires open to inspection I could decide what was usable and what needed to be replaced. I still made a like-new harness using the connector ends and some like-new wires complete with connectors from the old harness. I actually had a lot of the same size wire on hand with the correct code colors (or ones close to it).

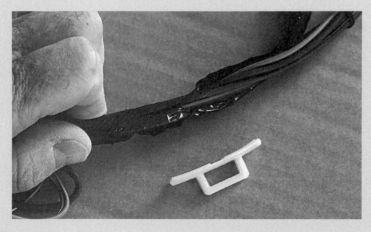

8. This harness clip just fell off and now we need to put it back in position. Remember, we are copying this harness exactly, or very close to it, so it will fit just like the factory one.

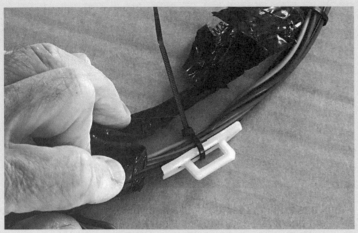

9. Place it back into position and run a wire tie around it so it stays in position. You really need to get this one tight or use several to make sure it doesn't move.

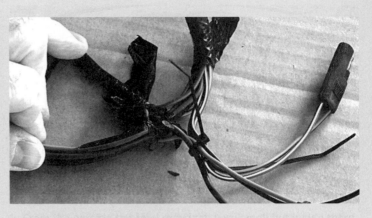

10. Here you can see that we have reached the main junction of this harness. We have removed the wrap, placed black wire ties in every position and we will now continue to take off the harness wrap. Note you can see a small nick I made as I was getting through to the center of this junction. It will be repaired before I go too much farther.

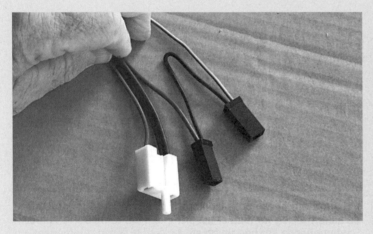

11. Here you can see a jumper wire from one connector to another. It is a ground wire so it is just being used to make a common ground for the circuit. This is the way to save money in wire by the OEM's. It might seem like they are only saving a couple of cents to a nickel at most. But multiply that times 500,000 vehicles and it is now $25,000 that can go to the bottom line. I left these wires alone as they were pristine. Only three small connectors were bad.

12. The wire without a connector is one of the three bad ones. To fix it I went through my old stuff and found a replacement connector that will work just perfectly here. It is off an old Mopar harness from some time ago.

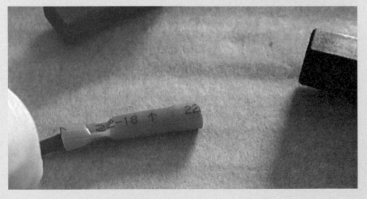

13. The first step was to place a crimp splice on the harness wire. Strip it and crimp it and you are good to go. I then placed shrink tubing on the next wire to be added before I spliced the like-new Mopar connector to the other side of this crimp splice.

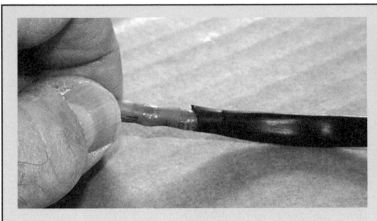

14. Once that was done, I was able to move the shrink tubing over the splice. It was heated and allowed to shrink.

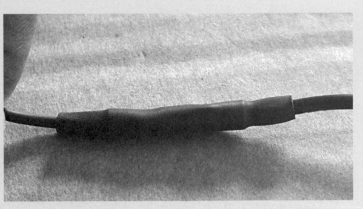

15. This is the completed splice ready to go. Nice and tight, strong and very professional looking.

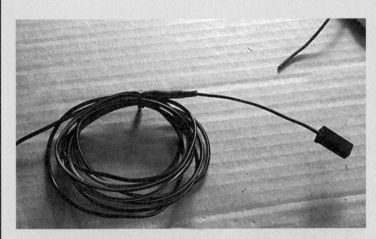

16. I had removed the entire wire from the harness so I was ready to replace it in the harness. To do this I laid it in place at the connector end and started pushing the wire through all the white wire ties. Where they were too tight I cut the wire tie off and then placed a new one over the bundle that now contained the black-with-red tracer wire.

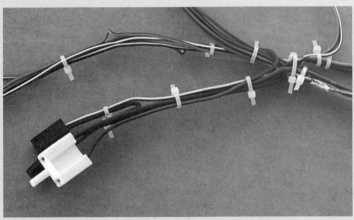

17. Here is a completed section, but until all of the repairs and changes are made we will not tape it. If you look closely you will see that some of these wires are jumpers to other connectors.

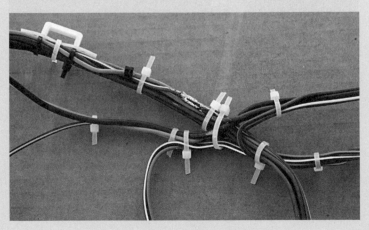

18. This is the major junction ready to tape—the nick in the wire was fixed and the solder joint has been copied. I'm sure the white vs. black wire ties now makes sense.

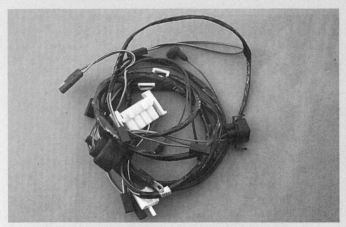

19. The harness is ready to be installed. It looks stock, all connections and clips are in place and it required only six splices and one solder joint. I guess this is now a rebuilt or remanufactured harness. No matter what you call it, it is now probably okay for another 30 years.

20. If starting at the alternator, decide where to run the wiring from the alternator to the central power buss (you'll also have to decide where that would be), so that power could go to the battery and the circuits running through the rest of the vehicle. It would have to go to the battery and the fuse panel as each circuit needs the protection of a fuse also.

21. If using a bulkhead connector, I recommend a Painless Performance 12-circuit-fuse center panel without wiring or a similar product from another supplier. Use a grommet to go through the firewall into the vehicle without causing leaks. Probably the best alternative is the Painless Performance part #3001 mentioned previously. But in this case split the grommet to run the wires through the firewall so they aren't stuck together. The last thing is to install the grommet and seal it with silicone. Because the owner wanted a stock look, he was given the System One catalog for Mopars.

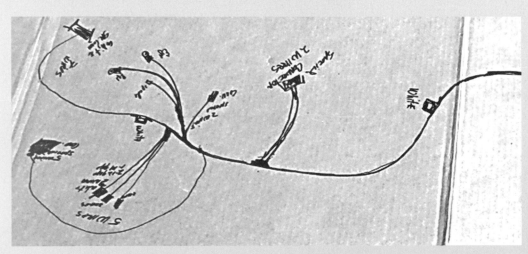

22. Another tip is to make your wiring diagram as you connect each circuit. It must include wire colors and locations as well as other notes where necessary. If you look you can see that they are all run together when they are a common circuit. Some have notes on them using masking tape and a felt tip pen. At some point I'll transfer this to paper and draw it a bit neater. You'll be glad you did years later when trying to isolate a problem.

SPARK PLUG WIRES

Wires take very high voltage and current from the coil and distributor then deliver it to a spark plug. Spark plug wires are a highly technical part of a daily driver as well as a hot rod, street rod or even a race car, so choosing the right one for your project may appear difficult. Plug wires come in a variety of core materials and insulators and there are many quality levels. If you have a basically stock engine, a standard set of aftermarket wires purchased from an auto parts supplier will probably get you by. But if your vehicle has headers, a custom engine, special belt drives, high-performance distributor or any other modification, you will be unhappy with standard spark plug wires.

There are three key issues with stock or less-expensive wires. First, if they have high resistance they will not get maximum power to the plug. Second, they may not always suppress RFI or radio frequency interference, which will cause static on your radio. Third, they may not suppress electromagnetic interference (EMI), which will disturb electronic systems, like the ECM or EFI. Performance aftermarket wires are designed to eliminate these problems.

When it comes to spark plug wires for a performance installation on a daily driver, street rod, hot rod or race car choose a wound copper wire core wire with silicon insulation and quality connectors. Here are some more facts on choosing high-performance spark plug wires.

Carbon Suppression Conductors

Carbon conductors are used in original equipment ignition wires by most vehicle manufacturers, and in the majority of stock replacement wires. This style of ignition wire is cheap to manufacture and generally provides good suppression for both RFI and EMI. The conductor usually consists of a substrate of fiberglass and/or Kevlar over which high-resistance conductive latex or silicone is coated, and functions because it can reduce spark current (using resistance) to provide suppression—a job it does well while the conductor lasts. Vehicle manufacturers treat ignition wires as service items to be replaced regularly, therefore limited life is never an issue. This type of conductor quickly fails (burns out) if a high-powered aftermarket ignition system is used.

EMI (Electromagnetic Interference)

EMI from spark plug wires can cause erroneous signals to be sent to engine management systems (electronic fuel injection and all computer engine functions) and other on-board electronic devices

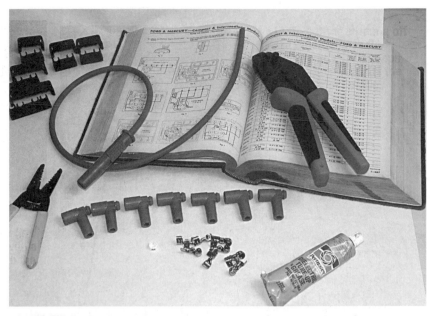

It is relatively simple to make your own high-performance plug wire subharness. These are important ignition components, but if you do not have a quality set of high-performance spark plug wires and a coil wire, all the rest of the equipment cannot help you. To do the job you need a shop manual, plug-wire set, crimping tool, distributor side terminal ends, distributor boots and Permatex dielectric grease.

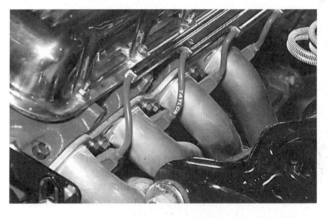

These plug wires deliver +20,000 volts to each plug as it fires in a 10:1 compression, wild cam Chevrolet 454 engine. With a big Holley carburetor delivering fuel, it needs this kind of powerful spark to burn everything in the cylinders.

used on both racing and production vehicles in the same manner as RFI (radio frequency interference) can cause unwanted signals to be heard on a radio. Engine running problems ranging from intermittent misses to a dramatic loss of power can result when engine management computers receive signals from sensors that have been altered by EMI signals emitted by spark plug wires. This problem is most noticeable on modern production vehicles used for commuting where virtually every function of the vehicle's drivetrain is managed by a computer.

For many reasons, the effect of EMI on engine management computers is never predicable and problems do become worse as sensors, connectors and wiring deteriorate and corrode over time. The

The quality of the terminal ends is also important. There are three types of terminal ends that will fit the 7mm silicon insulated plug wire on the right. Left is a stainless steel end with built-in spring, center is an aluminized steel end with two expansion collars, and right is plain brass. The terminal on the left is the highest quality, while the brass and other steel are typical of premade, decent quality low-dollar terminal ends found on standard sets.

This is the set of MSD plug wires selected for the Holley-equipped 351 cid engine we are building as part of this book. The engine is being built for high-performance street touring or class type drag racing. It should deliver about 400 hp and a similar amount of torque. I'll be using steel terminal ends as I fit the plug wires to the engine after the headers are installed for fitting.

Capacitor discharge ignitions (CDI) such as those from Accel, Crane, Holley, Jacobs, Mallory and MSD create sparks that are compressed (and intensified) into shorter duration and are designed to produce the extra spark energy needed by race and modified street engines that will reach higher rpm than stock engines. Some CDI ignitions incorporate multi-spark circuits to enable the engine to run smoother under 3,000 rpm. If you're upgrading to such a high-performance system, you'll need to make sure the wires can handle the extra output.

problem is often compounded if you replace the original ignition system with a high-output system but still use the stock plug wires. It is also important to make sure the wire terminal ends are of excellent quality to match high quality high performance plug wires.

Solid-Core Conductor Wires

Solid metal (copper, tin-plated copper and/or stainless steel) conductor wires are still used in some racing on carbureted engines, but they shouldn't be used at all as far as I'm concerned. They can cause all sorts of running problems if used on vehicles with electronic ignition, fuel injection or engine management or timing systems. Solid metal conductor wires cannot be suppressed to overcome EMI or RFI without the addition of current-reducing resistors at both ends of wires.

Low-Resistance Wires

The most popular conductor used in ignition wires destined for race and performance street engines are spiral conductors. Spiral conductors are constructed by winding fine wire around a core. Almost all replacement manufacturers use constructions that reduce production costs in an effort to offer ignition component marketers and mass-merchandisers cheaper prices than those of their competitors. Cost should not be the

determining factor in choosing which products go on your high-performance vehicle.

A high-output inductive ignition system is probably more appropriate than a CDI ignition system for late-model production engines (modified or not) because this type of ignition provides the longer duration spark needed by these engines that run super-lean fuel mixtures. High-output inductive ignition systems are currently available in the aftermarket from at least Accel, Crane, Holley, MSD, and a menu-driven direct ignition system is available from Electromotive.

Often, on production vehicles used on the street, replacing a tired ignition coil with a higher-output ignition coil like those from MSD, Nology, etc., can improve ignition performance, particularly under load at higher RPM. But, few coils can actually make horsepower; instead they simply release the horsepower your engine combination should have been producing all along.

Plug Wire Kits

Plug wires come in different sizes (total wire and insulation size in millimeters (mm). As you go bigger on the size you need to buy special distributor caps to hold the larger wires. I think it is easiest to buy a high-quality set of wires, cut them to the size you need, and crimp the connectors on the distributor cap end. MSD, Accel and Taylor all make great kits. I chose the 8.5mm MSD because we are also using an MSD billet distributor and their 6-AL ignition box. But the kit is only part of the process, you will need to determine the length, the loom system, etc. I recommend that you measure only after you have the headers on the engine, right before you install the engine. That will ensure you measure properly and do not leave yourself short. (Long is OK, since they can be routed or shortened.)

Spacing Looms—It is always best to keep plug wires from touching to prevent the possibility of high voltage jumping from one wire to another, but that sometimes can be avoided. Just make sure that any wires that cross are not in sequence in the firing order, which is what a good loom system is designed to do (as well as keep them from hot objects like the exhaust manifold or headers). That is the reason why looms must be custom made for a particular engine/header/head combination. Moroso, MSD, Spectre and Taylor and make good wire loom products. Just make sure the set you choose keeps plug wires off hot exhaust manifolds and headers on your V-8 engine. They say you can get a perfect fit using the adjustable mounting tabs (including SB-Chevy center bolt valve covers). Mounting blocks are made of a high dielectric material that prevents cross-firing between wires. They accommodate 7mm, 8mm, 8.5mm and 8.8mm wire sizes.

There are basically two types of crimping tools to attach the plug ends. One is supplied with the kit and the other is a professional tool designed for the task. The pro tool doesn't make financial sense if you are only going to make one or two sets of wires.

The other type of tool is a little primitive, but it is free and will work just fine for making up one set of plug wires. What you do is set the wire and terminal connector inside the tool, then place the other block into the first and then place it in a vise and tighten the vise until the tool is fully seated together. Then you take it out of the vise and check the quality of your work. I strongly recommend practicing on an old set of wires first.

Stripping the insulation to get to the core requires extreme care because it is so easy to cut into the

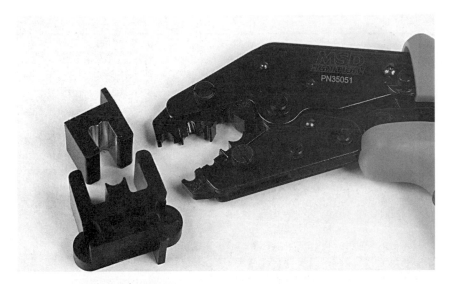

Here are the two crimping tools from MSD. The one on the left is a common one found at most auto parts stores. It is effective, but a little trickier to use and requires a table vise. It is included with the MSD plug-wire kit so it is free. The crimping pliers on the right are a high-quality tool that you should invest in if you are going to be making more than two sets of plug wires at some point. For greater ease of operation and quality results, the pliers are the way to go.

core, something you should avoid. Unless you have the right tool to do the job, it needs some practice. The MSD crimping pliers have a stripping function that makes it an easy job. The result is a well-stripped wire that gets you to the center core without cutting it. Traditional wire strippers can be used, but extreme care must be taken so you do not cut into the core itself.

Remember, the key to good plug wires is getting a quality kit to start with and then install them using a quality wire loom. You need to make sure the wire terminals you crimp are of good quality as well as the boots for the wires. In some cases they come with extra protection from heat by using boots that are made of a fiberglass-type material that reflects heat from the engine away from the plug wires itself near the plug. You might surprise yourself at just how easy and quick it is to complete your first quality plug wire and loom installation.

MSD'S PLUG WIRE KIT

You will need to know the firing order as well as each number cylinder to make the wire set. Ford numbers cylinders 1, 2, 3, 4 on the right passenger side, number 1 being the first on that side starting from the front and on the left/driver side are 5, 6, 7, 8, again starting from the front. GM and Chrysler number their cylinders with number 1 being the first at the front on the left side of the engine and 2 being the first at the front on the right side of the engine. They go back and forth to the back of the engine with number 8 being the last on the right side of the engine at the back. Keep this in mind as you install distributors and plug wires. All of the even numbers are on the right and odd on the left on GM and Chrysler V-8s.

You also need to know the rotation direction of the distributor rotor as it will help you set it in perfectly and to set the timing so the vehicle will start.

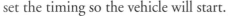

1. The MSD wire kit has the plug wire and cylinder numbers included. These are plastic numbered rings that you slip onto each plug wire as you build it. That way you will never slip the wrong plug wire on the wrong distributor post.

2. To begin, place a 1" length of wire through the tool, making sure you are in the stripper function, then squeeze the pliers, which will cut the wire to the depth of the inner core. Grab the cut end and twist it until it spins and you can remove it.

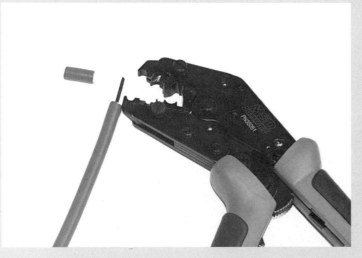

3. Once the wire is stripped and the core is clean, slip on the Pro Heat Guard sleeve and bend the 1" of inner core over tight to the end of the insulation and place the wire core against the terminal end you are going to crimp. Do not leave it so long it sticks out of the terminal end towards the top of the wire, as this can cause sparks to jump and bleed power from the plug.

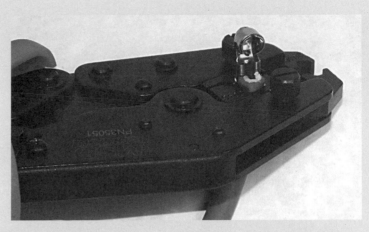

4. Once it is set correctly, slip it into the pliers with the open end of the terminal pointing to the round part where the pliers do the crimping (not where there is a point in the middle). Squeeze to tighten the jaws until the crimping tool closes completely and you can feel it release pressure and the ratchet gives way. The unit should look like this when it is correctly installed in the vise. Tighten the jaws until the tool is fully seated and release it. Both 7mm and 8.5mm wires are shown here with two different styles of terminal ends.

5. Regardless of which tool you select, this is how the completed crimp should look.

6. The next step is to stick some dielectric grease into the small end of the distributor boot and stick the terminal end of the wire into it.

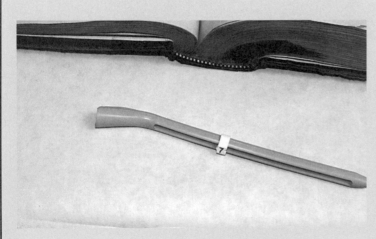

7. Getting that spark plug wire number onto the plug wire is a simple task using the supplied MSD tool. Slip the number ring onto the tool oriented the way you want it to read. (On number 8 that isn't an issue but on others such as 6 and 9 it is.) Be consistent in their orientation.

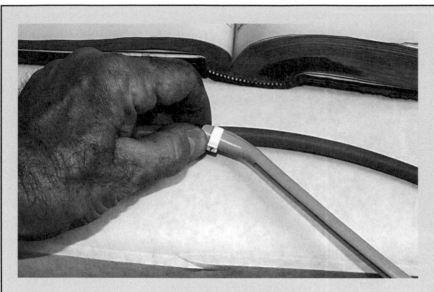

8. As you push the number ring down to the fat end of the tool it opens up the wire ring so it will slip over and pop right on. A very cool little tool and system.

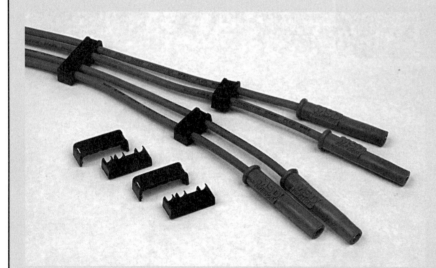

9. Here is one side of the set as they might look ready to install into a loom and the distributor itself. Once again, we will not know what the loom will be until the new engine is complete. The terminal ends, heat shield and numbers are not installed yet for the same reason.

10. This is the other end of spark plug wires, as they end up in the distributor cap. It is an MSD billet unit designed for the GM HEI (coil-in-cap) system. This eliminates the coil wire from the system and keeps everything a simple two-wire system hookup: only power and tachometer wires.

Nothing will deliver power and fuel economy like electronic fuel injection. Carburetors are still found on restored muscle cars, street rods, classics and on race cars that must run them according to class rules. But even old school hot rodders are retrofitting EFI systems to their project cars—it really is the way to go.

RETROFITTING ELECTRONIC FUEL INJECTION

There's little if any argument that electronic fuel injection offers many advantages over carburetion. It will make a performance or restored vehicle run with more power, more economically, across all levels of the power range. Many people shy away from using EFI, or converting their carbureted system, because they believe it is too complicated. While true that how an EFI system works is based on crunching a lot of sensor data, that doesn't mean you have to know all the details in order to install a kit. Today's aftermarket manufacturers have developed "plug n' play" complete systems, or even "piggyback" units that work with the stock ECM. Actually, the hardest part of installing an EFI system is bolting on the mechanical parts, installing a high pressure electric fuel pump, adding a return fuel line to your tank, if it doesn't already have one, and adding the oxygen sensor to the headers. Running and connecting the wiring is the easy part.

The engine control unit is the most critical element in an EFI system. There are many brands available, and one of the most popular is the Fuel Air Spark Technology (F.A.S.T.) system developed by COMP Cams in Memphis, TN. Although many muscle car owners shy away from EFI to maintain originality, there are throttle body injection kits that will retain that stock look. Although TBI looks closest to stock because it is a single unit mounted in place of the carburetor (and usually partially hidden by the air cleaner), it does not match the power and fuel economy of a port-injected EFI system.

In addition to a power source and a good ground, the ECU needs the following information to adjust air, fuel and timing to deliver maximum power and economy for the given driving situation:

• Engine Temperature
• Ambient Air Temperature
• Throttle Position Sensor (TPS).
• Engine Vacuum (Manifold Absolute Pressure or MAP sensor)
• Exhaust Oxygen Content (O_2 sensor)
• Engine Timing (Spark Advance)
• Engine rpm
• Engine Idle Speed (IAS sensor)

The mechanical parts may differ in the size of the intake manifold, fuel rails, injectors and throttle body, but otherwise they will be similar for all EFI systems.

The F.A.S.T. kit has just 8 connections (plus the fuel injector for each cylinder on a multiport injection system),

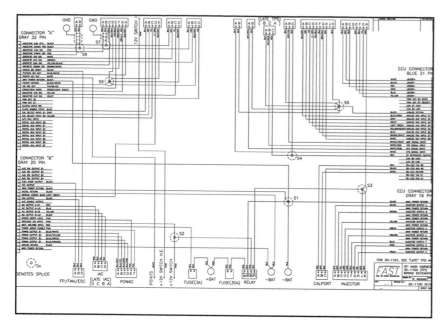

The connections for the components are made with weather pack connectors that keep the wiring dry regardless of the weather. F.A.S.T. also includes a very good wiring schematic that shows you which wires and connectors go where. The injectors are all numbered for the type of engine you have. Just match them up and make sure they match the engine's firing order. You can install the MAP sensor on the firewall behind the intake manifold to get a direct vacuum port connection. Although we used plastic tie wraps it can also be bolted in place. It is plastic and weighs next to nothing.

We installed the engine temperature sensor, just as you would install a temperature sender for the instrument panel warning light or a gauge. The F.A.S.T. intake manifold is machined for the injectors and sensors.

We went directly to the dyno for a tuning run and then to wring it out. After the tuning run, with David sitting in the seat setting fuel maps, we took the rpm up to the high end. At peak 6,000 rpm, the carb and EFI were the same because the carb had been tuned that way. But the big difference was throughout the rest of the powerband, where the EFI produced more horsepower and torque at every point. What's better, it did so with better fuel economy.

plus the power lead and a ground. You will also need a 2" hole saw for this project.

If the ECU is the brain, then the throttle body is the heart of a multiport EFI system. In this case it has four throttle bores and plates and passes the correct amount of air upon demand from the driver's right foot. You have to connect the throttle cable or linkage to it. The timed plate opening is achieved by the linkage between the front and rear bores. It allows the engine to gain rpm quickly on demand and eliminates lean conditions when accelerating fast. The throttle body assembly mounts on a carburetor manifold just the same and installs in a couple of minutes, but there are no fuel lines to connect. The TPS is always on one side of the throttle body because it needs to measure how far the throttle is pushed down.

On this car we had to replace the 4-bbl carburetor intake manifold with one that had bungs for the direct port injectors and fuel rails. We also had to add a complete fuel pump system (w/regulator), mount the MAP and ECU, install the engine temperature and air temperature sensors, add the throttle body (it includes the throttle position sensor) and route and connect the wiring harness to all 8 of the connections, plus injectors. Each connection is made with male and female connectors, which makes the job much easier.

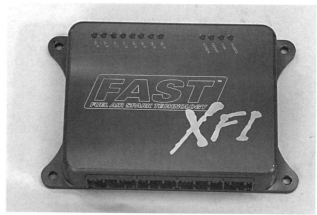

This is the F.A.S.T. electronic control unit (ECU), the brain of the F.A.S.T. aftermarket EFI system. After mounting, all you need to do is plug the appropriate connectors in and give it a fuel map to follow.

INSTALLING A F.A.S.T. KIT

We had the opportunity to do a complete install at F.A.S.T. headquarters, part of COMP Cams in Memphis, TN. It took us about 3 hours more than it should, because they wanted the ECU in the engine compartment, not behind the kick panels or under the dash where it would normally be placed for maximum protection. Turns out the Mustang is a show car for F.A.S.T., and they wanted the entire product line visible underhood for demonstration purposes. But, even with that we were done in less than a full day.

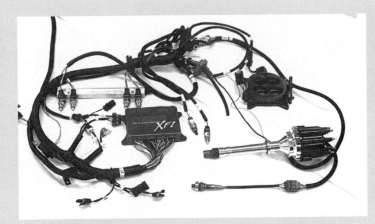

1. This is the complete F.A.S.T. wiring harness with almost every component connected. From right to left are: the oxygen sensor, rpm and distributor timing, the throttle body, which includes the throttle position sensor, MAP sensor, engine temperature sensor, air temperature sensor, 4 of the 8 injectors and the ECU connected to the harness.

2. We are retrofitting the F.A.S.T. kit on a 1969 Mustang convertible with a strong rebuilt small block.

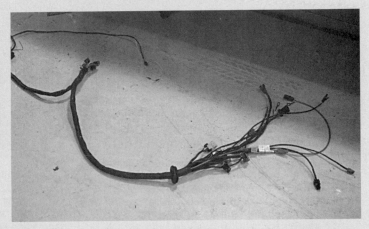

3. The modified wiring harness has far more weather protection than normal because it has to go through the wheelwell, where It will be exposed to road grime and water. If you compare this harness to the standard one you will see the changes that needed to be made to move the ECU outside to the fenderwell. All of the wires and connectors on the right side of the photo will need to be inside the engine compartment and those on the left, inside the car. Remaking the harness isn't something I would recommend, even though the guys at F.A.S.T. did it in about 3 hours. But then again, that is what they do all day long.

4. Mounting the ECU on the inner fenderwell makes it easy to see the operational lights, but it is more exposed to heat and the elements. Since this is a show car, it was okay, but for the street you'd want to mount it inside the car, usually somewhere in the passenger footwell. To mount it here, we had to make sure there was room to drill a 2" hole under the ECU, and that there are no exhaust pipes anywhere near it.

5. We cut a hole in the fenderwell to bring the wiring into the engine compartment. In a regular installation, all you do is cut a hole in the lower firewall so you can bring the wires into the passenger compartment. Usually the ECU is mounted in a kick panel or under the dash. We tried to drill from the inside of the engine compartment, but ended up taking off the wheel and drilling through that way. A small pilot hole was drilled first so we could see if we were going to hit anything when drilling the 2" hole.

6. All of these connectors must be fed through the 2" hole we drilled so they can be connected to the components inside the engine bay. These would normally be coming through the firewall or from the backside of the fenderwell. Notice the grommet to protect the wires from chafing on the sharp metal of the hole we drilled.

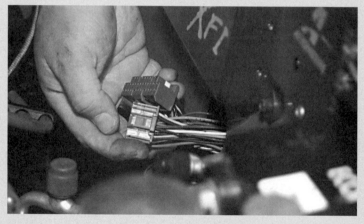

7. Although all these wires look like too much for the 2" hole, they will fit and still have room for the grommet. Just push them through the hole one at a time and then push the grommet into place on the fenderwell.

8. Although there are a bunch of wires to deal with, their number is reduced because they terminate into Weather Pac connectors. Here are the two main connectors for the ECU coming through the hole. They cannot be installed first since the ECU is right above this hole and there are still more connectors to pull through.

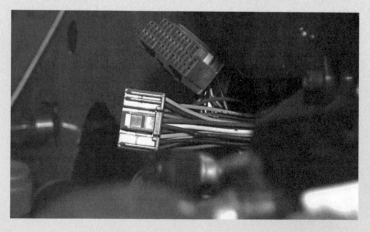

9. Here you can see some of the connectors coming through the hole to the engine compartment. In a regular installation none of this would be done as everything would be connected inside the engine compartment and then the rest would go into the passenger compartment.

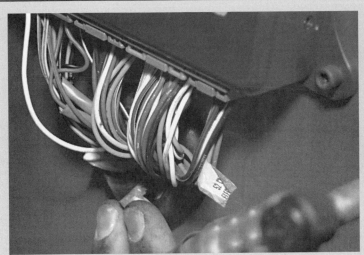

10. When all connectors are through the hole, make the connections to the ECU and then you can get back to a standard installation.

11. This is what the harness looks like in the wheelwell area. Never run a wire through a metal hole without a grommet, which is shown here. The grommet keeps water outside of the engine bay and keeps the harness safe. Add some silicone sealer at the parting line to secure the grommet and make it water resistant.

12. The throttle body with throttle position sensor allows air into the engine based on how much the gas pedal is pushed down by the driver. Aside from the electronic connection, however, the mechanics of the unit are like the throttle plates and bores of a carburetor. It controls how much air is allowed to come into the engine.

13. This is the intake manifold with the throttle body in place where a carburetor would usually reside. A fuel pressure regulator is on the front with a pressure gauge, injectors and fuel rails are in place and the entire assembly is almost ready to install. The engine-temperature sensor can be installed on either side of the water inlet. The engine temperature sensor must be installed in the "wet" side of the manifold, the same place where you'd install the sending unit for an engine temperature gauge for the dash. On this manifold there are two locations in front. If you do not use both (sensor and gauge), remember to plug the open hole. Use a sealer for both the gauge or plug and the F.A.S.T. sensor. (To those sharp-eyed enthusiasts that spotted this was a Chevy manifold instead of a Ford, good job! We were just using it to make the point, it didn't go on the Mustang.)

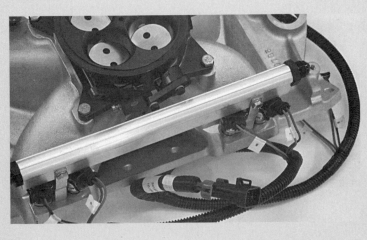

14. In this view you can see the throttle plate linkage. It is staged so the primaries open quicker than the secondaries. The fuel rail and 4 of the 8 injectors and their connectors are already connected. Each injector has a number, so make sure the number matches the cylinder it attaches to.

15. One gasket, 4 bolts and the throttle body is ready to mount in the place a carburetor would normally sit. The manifold change took about 30 minutes after the radiator was drained. Just be sure to clean off all gasket material and use new gaskets. I always use silicone sealer at the ends of the manifold because the rubber and cork gaskets always leak oil.

16. This is the Manifold Absolute Pressure sensor (MAP) that measures intake manifold vacuum or pressure to let the ECU know what is going on with the engine. This tells the ECU how much load the engine is trying to meet based upon the vacuum or lack of it. It works the same with a pressurized system (turbo or supercharger). It keeps the ECU happy and able to make smart decisions. It is generally mounted behind the intake manifold on the firewall so it gets a straight shot at a vacuum source from the intake.

17. The engine temperature sender on the left goes into one of the wet openings in the intake manifold. Just be sure to plug and seal it with a good pipe thread sealer. The air temperature sender goes into the air cleaner or the dry side (vacuum port opening) of the intake manifold. The manifold keeps it out of harm's way while the air cleaner location will allow it to get a reading closer to the ambient temp.

18. The throttle position sensor snaps right in place. Remember this is an important data source for the EFI as it lets the ECU know what you want

19. This is a Chevy distributor as modified by F.A.S.T. They also have Ford and Mopar units available. The system will also work with multiple spark units such as the MSD 6-AL series. The F.A.S.T. distributor, like any electronic distributor, plugs right into the wiring harness. Using the F.A.S.T. distributor eliminates the need to change the connector to fit the "weather pack" connector like the one on the female side in the harness.

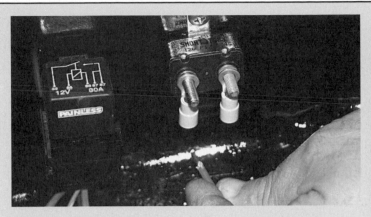

20. If you want to protect the wiring as well as get back racing fast, install a circuit breaker of the same amperage. That way if it pops the circuit open, when it cools down it will close the circuit. At least this might get you home or down the racetrack to the pits.

21. All of that pump and filter equipment goes in the back of the car under the fuel tank or fuel cell and it draws a lot of current. This older system was already in the Mustang and is shown to demonstrate where things go. This pump was plenty to fire up the car with the new F.A.S.T. system.

22. You will need to have an O$_2$ sensor mounted to your exhaust. If the collector is 3" in diameter, mount it 6" down the collector. Also make sure it is mounted to the top half of the collector or exhaust pipe. The oxygen sensor is installed into the exhaust system on your headers or the exhaust pipe close to the spot where all cylinders dump into the exhaust. It should not be on just one pipe from the headers as this will cause the system to think the average O$_2$ is for 4 cylinders when it is actually coming from one. That could cause big problems.

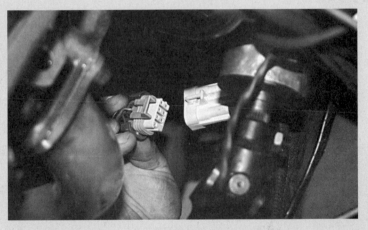

23. Once the bung is welded on, the hole is drilled in the pipe and the O$_2$ sensor is installed using anti-seize, run the connector to its harness connector and plug it in.

24. This is the single wire from the MSD box to the F.A.S.T. harness that delivers the trigger signal to the spark plugs and guide injector timing. All complicated technically but for us it was a single spade connector.

25. This connector is what powers the injectors and tells them when to squirt. It implements a signal from the ECU to keep each injector at a predetermined bandwidth (how long it stays open).

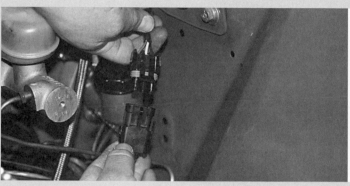

26. This is where the air temperature sensor connects to the system wiring.

27. In this view you get a good look at the braided fuel hoses running from the right side across the firewall. Since this is a high-horsepower application, F.A.S.T. used a Y fitting out of the main fuel line at the rear of the engine and then a hose leading up to each fuel rail. This parallel system delivers the maximum amount of fuel to each injector. (That's right, the fuel goes into the rear of the fuel rails.) From there it gets injected into the cylinders or bypassed back to the tank by the regulator.

28. In this photo of the fuel regulator you can see the mounting to the manifold and the relationship to the fuel rails. One hose leads out of the front of the rails to either side of the fuel pressure regulator. Run one line from the front of each rail to each side of the regulator. The fuel regulator works by sending or retaining fuel to the return line until the pressure is correct so head pressure is maintained. The fuel return lines goes into the air space below the manifold back to the tank.

29. This is the fuel pump and filter system. It includes the pump, cleanable filters, regulator, relay, circuit breaker, wiring and wire ties. All fittings are included, but no lines are included due to the variables in many possible installations.

30. For additional peace of mind, wire in an oil pressure switch set for the minimum oil pressure so if the engine quits running or loses oil, it will shut off the pump. Wire it to any existing oil idiot light circuit if there is one, or create your own. It must open the circuit when there is no oil pressure.

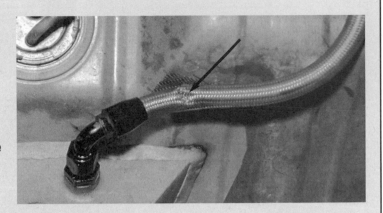

31. Note the road/rock damage to the suction line going to the pump. Good thing it was braided high-pressure line or the car would had serious trouble.

32. It is easy to tune the system once it is running. Just connect your laptop with the software that goes with the system and follow the instructions and screen prompts. F.A.S.T. has a tech line to help answer all of your questions.

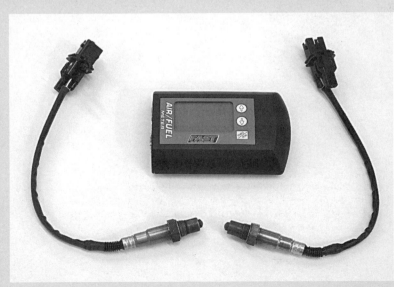

33. This unit allows you to tune while you drive. It can work with the F.A.S.T. EFI system giving you even more drivability and fun.

(TBI) THROTTLE BODY INJECTION THE GREAT ALTERNATIVE

Throttle body injection may not be as precise as direct port fuel injection, but it is still more accurate and efficient than carburetion, and it is an easier replacement kit than EFI. One of the more popular systems is Holley's Commander 950 Pro, Total Engine Management System, which we installed on a small-block Ford 351. The difference between the Holley TBI system and a multipoint port system, is that TBI sprays the fuel directly into the airstream above the throttle blades of the throttle body, while EFI sprays it directly into the port. TBI operates the same way as a carburetor, but it is much more precise. This system adjusts for air temperature, altitude, engine load and also allows you to tailor fuel curves for the specific needs of your engine. It doesn't need a direct-port EFI specific intake manifold, ultra-high fuel pressure, and any other of the special requirements a direct-port, race-type EFI system needs. So TBI systems are therefore less expensive, and the performance improvement (more power throughout the RPM range and better fuel economy) make it very attractive, even to old school hot rodders. So aside from where and how the fuel is injected, both EFI and TBI are similar in principle, and use the same engine management sensors. Because of this we will not review the entire installation process for the Commander 950 Pro TMS kit, but we will review the steps that are mechanical and install all the sensors. The Ford engine we installed the kit on was rebuilt with the following:

• overbored 0.030"
• COMP Cams High Energy hydraulic 278° shaft with 0.500" lift, 114° lobe centerline
• Holley 17° aluminum cylinder heads with angled spark plugs, screw-in rocker arm studs, 2.02"I/1.60"E valves.
• Weiand Stealth 351W intake manifold

1. It is always best to spread out all the parts so you can identify them and visualize exactly what you need to do.

2. What makes the throttle body injection system so attractive for the hot rodder is that it all fits under a 4-bbl air cleaner, it uses a stock or aftermarket 4-bbl intake manifold without any modifications, it will outperform a carburetor in overall performance and it is reasonable in price.

3. The other side of the unit shows the injector wires and the air-temperature probe sensor (the black tube). You can also see the small plenum under the injectors that allows the mixture to be drawn into the cylinders that need it the most at any given moment.

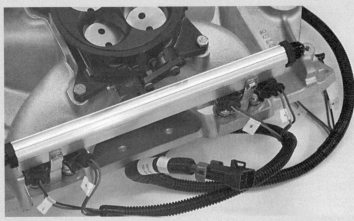

4. Compared to multi-point injection, the TBI system uses far fewer components. No fuel rails, 4 instead of 8 injectors, stock instead of custom intake manifolds. In the electronics, both systems are about the same and allow you to select fuel maps to tailor the system to your engine's power level and requirements. For cruising, the TBI fuel injection system is perfect as it will help you get to the 20 mpg mark if you install the rest of the mileage combination.

5. Above the throttle body are 4 constant-flow fuel injectors. They can meter out precise amounts of fuel based upon the input the ECU receives from the sensors. From the top you can see how each of the injectors is directly above a throttle blade. That gives it a direct shot into the plenum just as a carburetor would have. The difference is that the ECU can adjust for driving conditions like altitude, engine temperature, throttle position, cold-start fuel enrichment, high ambient air temperature and humidity, unlike a carburetor.

6. Although still not as precise as direct-port EFI, the TBI system is still a major improvement over carburetion. The TBI is a giant step forward in fuel and engine management. From the left to the right, you can also see the throttle position sensor, then the idle air control motor and in front, the throttle linkage and finally the wires leading to the injectors from the injector harness connector. That black tube sticking up is the air-temperature sensor.

7. The throttle body base is just like the base plate on a carburetor. The throttle linkage hooks up with a little adjustment available to tailor it to the engine.

8. This is the rear of the unit, with the fuel inlet hose nipple on the left, the fuel pressure regulator in the middle and the fuel return hose nipple on the right. Uniting everything into a simple and easy-to-understand throttle body assembly is why this system works so well for many early-model performance engines.

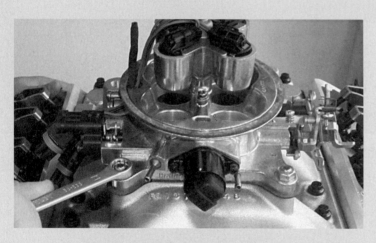

9. This an overall view of the front of the unit. The idle air control motor goes to the front of the intake manifold. The idle air control motor calculates actual idle speed vs. a graph of what is most desirable for the system based upon air temperature and air flow. This is a stepped motor that can adjust it up and down to match the desired graph level.

10. This is the MAP (manifold absolute pressure) sensor. It should be mounted on the firewall right behind the engine. One vacuum connection and one connection from the computer and it is good to go. It tells the ECU how much load it is being asked to deliver. For an old school rodder, it is basically an electronic vacuum gauge. But this is one of the most important sensors in the TBI system.

11. This is the time to install the EFI sensors. The temperature sensor is in the front of the engine to the right of where the MSD distributor will call home. The air-temperature sender is in the bundle of wires going to the tip of the injectors. It is a black tube with wires that go down to a connector at the base of the injector body. The MAP sensor will be mounted directly behind the engine on the firewall and all the other connections are already on the TBI unit so you just have to run the harness to the engine and inside to the ECU. Don't forget you must install an O_2 sensor in the exhaust.

12. The kit also includes the mounting for the transmission pressure control cable so you do not need to go anywhere to buy anything; except the fuel line if you do not have a return line for fuel to flow back to the tank or you want a bigger line on the inlet pressure side of the fuel regulator.

13. This is the brain behind the Holley Commander TBI system. It takes all the information fed to it and makes decisions regarding how to maximize fuel economy, performance and drivability based on current driving conditions. Mount it inside the car behind a kick panel (driver's side or under the dash).

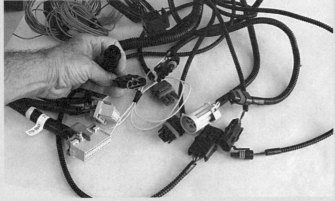

14. This pile of wires looks complicated, but is a simple job to install. Just 7 major sensors plus the square connectors to the ECU power supply, relay and that's just about it. Install all the mechanical stuff on the engine, the fuel pump, fuel lines and other bits. Then you can connect the wiring harness to everything in the engine compartment before drilling a hole into the firewall so you can send the interior cables inside. The stuff in my hand goes into the interior of the car along with the blue connectors for the ECU just below my hand.

15. The kit also comes with a plastic air cleaner spacer to raise it from the TBI unit. A new gasket for below the TBI unit and some new vacuum hose is also included.

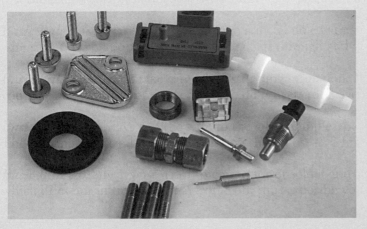

16. The mechanical bits also include (from left to right in circle); firewall grommet, fuel pump block-off plate (w/bolts), MAP sensor, plastic fuel filter to go between the tank and the electric fuel pump, engine temperature sensor, throttle return spring, TBI mounting studs, (middle in circle left to right) AIR pipe connector adapter, O_2 sensor bung to be welded on exhaust, TBI system relay, air pickup.

17. The TBI system O_2 sensor is included and the sensor is already treated to anti-seize compound. Make sure you weld the sensor bung into the top half of the exhaust pipe or header collector.

18. The high-pressure electric fuel pump gets covered with the black foam in the background to eliminate vibration and then mounts with the two insulated clamps included in the kit. Remember to install the NPT nipples on each end and run a rubber line from the tank to the plastic filter before running the next one from the filter to the pump. Use quality clamps for all connections. If you want to go for AN fittings and lines, remove all the NPT nipples and replace with the correct size AN fittings. If your tank does not have a return line, install one. Going from the front to the back be sure to use either hard lines or braided stainless mesh covered hose for protection.

19. As with the EFI system in the last section, you can use oil pressure to shut-off the fuel pump using this kind of switch, sometimes called a Hobbs switch. The oil pressure keeps the circuit closed (this switch is 50 psi, but you will need just 5 psi) as long as the engine is running. But if the engine loses oil pressure it will kill power to the fuel pump, which is safe in an accident situation. This is a simple ground wire that is connected to both and connected to each side of the spade blades on the pressure switch. This would be useful for arming fuel enrichment systems on supercharged or turbo motors also, not just to shut off a fuel pump.

20. This big filter should be mounted along the frame rail or in the back of the chassis before sending the fuel to the TBI unit. When it comes to injection systems there is no such thing as keeping the fuel too clean.

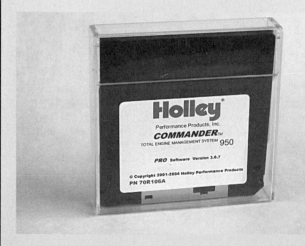

21. This is the programming software for the ECU. It already has some base fuel maps (probably just what you need for general driving) but it has plenty more as well as all the instructions you will need. If your computer doesn't handle floppies, Holley will send you a CD if you need it. It is also available online.

22. This is the bottom of the throttle body for the EFI system; the blades are closed.

23. With the throttle at full WOT the blades should be absolutely vertical and the only thing to slow down airflow should be the thickness of the blades and the screws that hold them in. On this throttle body they are not opening 100%.

24. I found the culprit to be this black throttle stop. It doesn't allow the throttles to open fully. (My finger is pointing at it.)

25. All it takes to fix this is to grind off a bit of the black piece so it can travel another 1/4". I can also grind it off the aluminum piece where the throttle stop hits. Either one or a combination of both will fix the problem.

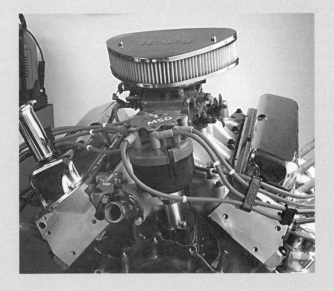

26. Now is the time to start it up, drive it on the starting fuel and ignition map and then bring it back home for more tuning. You will now know what needs adjusting and the programming will help you through the process. Once you have it where you like it the real fun begins. If you have problems call the 800 help line that appears on the instruction manual. From now on your vehicle will run better, start up easier, deliver far better fuel economy and make more power from idle to max rpm.

Chapter 11
Wiring a Custom Instrument Panel

If you want to have a custom dash like this sitting in front of you as you drive your hot rod, you need to have a plan and implement it perfectly. The beauty of a custom dash is incredible, but the work to make it look like this is really easy; it is the planning that is key.

The instruments in a car are just as important to the proper operation of a vehicle as the engine or transmission. You need to know the electrical system is working, that the engine temperature is within limits, how fast you are going, your engine speed (rpm), how much fuel you have, oil pressure, etc. All this information is important, especially if breaking in a new engine or operating the car in extreme conditions.

This chapter shows how to install all the gauges needed and how to wire and harness them for a professional custom appearance in any kind of vehicle. Most muscle cars, classics and restorations keep the panel looking stock, but with a street rod or race car, anything goes. If you are rewiring the vehicle, you can get an OEM or aftermarket harness, take the old one out and install the new one. There are also suppliers that will take your OEM classic instrument panel and restore it to like-new condition.

When you are starting from scratch that isn't an option, but you do have a wide variety of choice from modifying the stock panel to building a completely new high tech panel. In the case of our pro street truck project detailed in Chapter 8, the owners selected a billet aluminum instrument cluster insert from Billet Specialties and gauges from Stewart Warner.

The Billet Specialties panel consists of the speedometer set in the center with 4 smaller gauges clustered around it. We test fitted it into the stock nacelle to make sure it fit before installing gauges and wiring. We were going to use direct-reading water temperature and oil pressure gauges. That type of gauge has a direct connection between the engine in the

form of a sender cable for water temperature and an oil line for the oil pressure.

We wanted to create a quick-change panel by using a 9-pin connector between the panel wires and those under the dash. It would have been easier if the oil pressure and water temperature gauges used electrical senders instead of being direct reading, but we had plans to cope with them also. We still installed everything to a master connector, except the oil pressure line, which comes from the engine bay into the cabin of the truck under pressure and the water temperature wire. A professional look was created by using a flexible stainless-steel braided hose through the firewall and under the dash, where it is connected to the oil gauge. The water temperature has what looks like a number 8 black cable coming from the engine to the gauge, but it is a special assembly. (It is all preconnected from the factory and cannot be taken apart.)

Stewart Warner has been providing gauges for hot rods since the industry began in the '40s. For the pro street Ford, we selected the following gauges from their Maximum Performance line:

- Tachometer with Shift Light
- Electronic Speedometer
- Volt Meter
- Fuel Gauge
- Water Temperature Gauge
- Oil Pressure Gauge
- Electronic Sender Adapter for the electronic speedometer

1. By laying the components out on a clean piece of cardboard you can visualize the installation and have a nice work space that you can set all the little nuts and connectors where they will not fall and get lost. Everything is very light, so weight isn't a problem. We worked on cardboard that was placed over a trash can and against a wall so we had the splash back to keep from losing any small parts.

2. We installed the speedometer first. It goes in the center and is a bit more complicated to bolt in. Each gauge must be square so that the name Maximum Performance is aligned. We started by having the speedometer's #60 line aimed at the center little hole at the top of the billet insert. Install the speedometer in the center of the dash insert. It is held in position by three L-shaped brackets that hold it tight to the instrument cluster. Install it first and make sure it is straight in the front before you go to the next one. The other gauges are held in place by brackets that go across the backs of the gauge and down to the insert itself.

3. All of the other gauges were on the same improvised worktable along with the small tools we were using. The instructions for each, as well as the fasteners and connectors, were alongside each gauge. This made the installation easy because it was organized and everything was in easy reach.

4. After all the instruments were mounted, we turned the insert around before starting the wiring. If people at a show take a look inside your vehicle they will look at the dash first, so make sure each is lined up perfectly. When all gauges have the names lined up and are square to the top of the insert you are ready to tighten everything once more and continue. If you look at the top, you will see three small holes. The center one is for the high beam, with the turn signals on either side. Those little holes will be filled with small LED's, two green and one red.

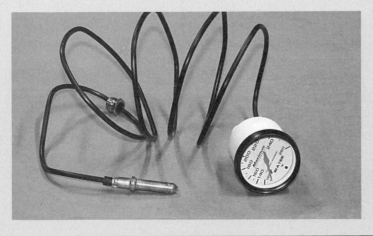

5. In this photo you can see the entire water gauge. The gauge gets mounted into the insert from the front and the cable and sensor go through the insert into the firewall, to be mounted into the intake manifold's wet side. It is easy to see why a street rod would be better off with an electrical sender, not a direct reading gauge.

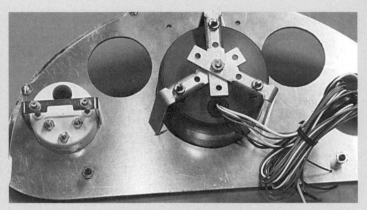

6. Use care to make a good seal when installing the sensor into the intake manifold to prevent coolant leaks. The fitting included with the kit is designed for this application as it has a sealing bevel inside it, so do not just grab another fitting of the same size even if it is handy. We installed the adapter fitting with Teflon tape and thread sealant on the sensor.

7. The speedometer is mounted with three L brackets and the smaller gauges have a U-shaped bracket. The small gauge is the voltmeter and the three posts on the bottom backside are for electrical connections, while the backside hole is for the lightbulb.

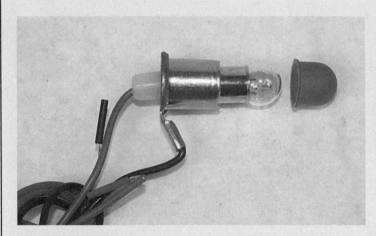

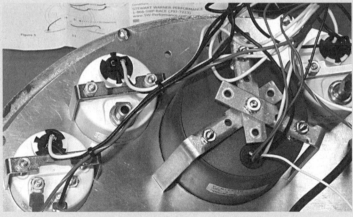

8. To select the color of the lighting you want in your dash, choose either red or green and slip the little bulb covers of your color choice over each bulb in the dash. As long as the lightbulb stays working it will be that color. But when it burns out, you will need to get another bulb cover of the correct choice from Stewart Warner. You can also leave them off and have white light.

9. Each gauge comes with a light socket and bulb to illuminate the gauge when the headlights are turned on. These bulb sockets all have just two wires, black (ground) and white (power). You will make your first harness on this installation.

10. Gather up all of the back wires from each of the gauges and join them into a single cable in the center of the insert. This will include all bulb black leads and the same with the white leads from the bulbs. Make sure you also include the two internal grounds for the fuel level and the voltmeter as well.

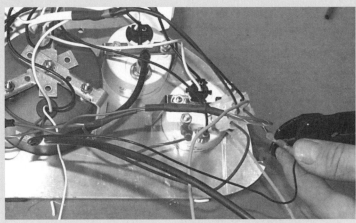

11. We included the ground and power wire of the speedometer into the harness we are building for the insert gauges as well.

12. On the voltmeter and fuel level gauges, you will find a red with a white tracer wire that gets connected to a post on the back of each electric gauge. These are the power wires for the internal mechanisms of the voltmeter and fuel gauge. Form them into a single cable at the center of the insert. You can see the wires being bundled and also where the single red/white wire goes into the female insert connector. The grounds for these two gauges are black and have already been bundled into the ground lead for the harness.

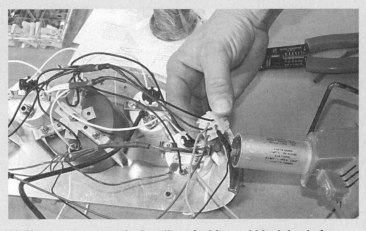

13. Here you can see the bundles of white and black leads from the instrument lights. The gauge power wires are also bundled together. Although everything still looks like a mass of wire spaghetti, it will come into an orderly form fast from this point. To connect several black ground wires into a single larger wire we use a crimp connector that is usually the female end of a pin connector. Since it is much larger on one side, so the pin can get in, we were able to insert all of the individual ground wires into the big end and make everything look neat. They are visible on the left middle side of the photo.

14. The next step is to bundle the wires with wire ties from each instrument to the next. Do it on each side and into the middle. On the voltmeter you can see the power and ground wire. The next step on this gauge is to find the voltmeter wire from the Painless Performance kit and attach it to the empty remaining post. In our case, we cut a 18" section from the Painless wire and installed it into the gauge and then the female connector so we could have a quick disconnect dash. Do the same on the fuel gauge as it also has a wire from the main Painless harness. You can see the harness is starting to take shape.

15. Once you have both sides tied together, you can tape over the bundles with electrical tape or run protective PowerBraid as supplied by Painless or other similar products. For us, electrical tape was the best solution. The wires are now neat and orderly, and all it took was about an extra 30 minutes of our time. Do not cover the power wires yet, even if you have to wait a bit before you place tape over the wire ties.

16. You can see how neat the harness looks and how the harness comes from each side of the insert to the middle. We could not have done this job with the insert attached to the dashboard. That is why we built the instrument cluster to be connected to the main harness using a multi-pin connector. (Male on the dash side and female on the insert side—even though it doesn't make any difference.) This is the female side of the connector.

17. Looking closely, you can see how we routed the wires into groupings before we actually bundled them into a harness. The lower gauge is for fuel level and the upper gauge is water temperature.

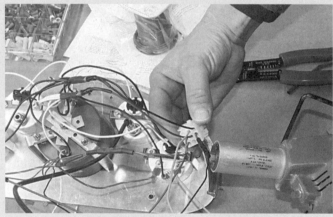

18. We used a lot of shrink tubing in this project because it supports crimp connectors, is very tough and simply looks great. We used both butane heat from a mini-gun and a heat gun to shrink it. We used a 3-to-1 shrink ratio with glue inside. (3/8", 1/2", 5/8" and 1" sizes).

19. The last step for the insert is to add the directional signal indicating lights and the high beam indicator. These go in the three small holes at the top of the dash insert above the speedometer. They just slip in from the front and are held in place by a small metal nut on the back.

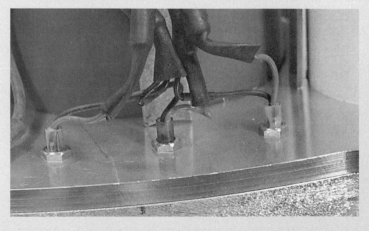

20. The lights are actually tiny LEDs that are available from places like Radio Shack. They have a power wire and a ground wire, so connect the power wire to what you want it to indicate. There will be leads from the main harness so once again cut an 18" length off and install them in the female insert connector on the insert side and on the underdash side. We used the red for power and they went to the matching wire from the Painless kit that was underneath the dashboard. Do them one at a time and then do the same from the matching pins on the female side and splice them into the correct LED power wires. (In this case they were the left-turn signal indicator (green), high-beam lights indicator (red) and right-turn signal indicator (green) in that order.

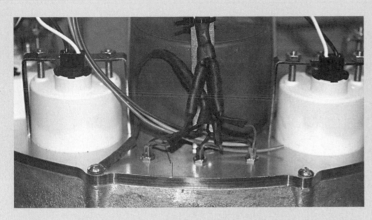

21. Just run the three ground wires (it doesn't matter which you use but we used black for continuity) to a single common ground wire and attach it to the dash insert which will be grounded to the chassis. (It is under the screw on the front left of the insert.) Now that this is completed you can tape over your bundles in the back.

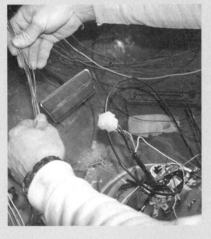

23. If you do the female side first you must chase down all the wires from the Painless kit and then add the pins to the appropriate matching wires and insert them into the male connector by color matching wires to the female side. Here you can see Pete McDonald with the dash insert in his lap as he locates and measures the wires for the male connector.

22. Next connect each lead from the harness to our female master connector as it is the key to creating a quick-change dash insert. Crimp each wire to a female pin connector and then stick them into a pin location in the female connector. Six pin locations were filled and ready to plug into the male connector: right turn, left turn, high beams, voltmeter, fuel gauge and speedometer power. By the time we cut and inserted all wires from the insert to the connector, we still had 18" of harness free-play between the insert and the connector. The white wire not in the harness is for the electronic speedometer. It is the calibration and trip mileage reset wire that gets mounted under the dash. Note: If you have already done the male connector in the dash, make sure you draw a diagram so you can match it with the female. Not doing it right can be fixed, but it is time consuming and a real pain. You can also connect the male and female connectors and then match colors on the pins, if you have used the wire from the kit for both sides.

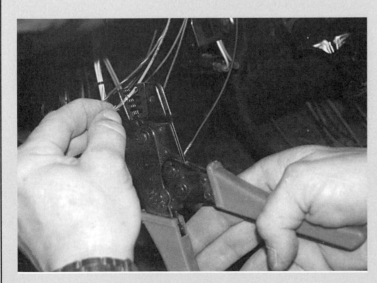

24. These are the crimping pliers to use when crimping the pins to the wires for the male and female connectors. Pete is crimping the fuel gauge wire from the Painless harness. Only five more wires to go. If you do not have these crimping pliers you can use a regular crimper if you are careful or you can use needle-nose pliers if you crimp the wires tightly.

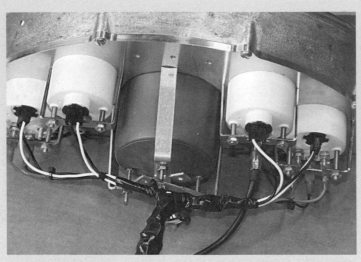

25. This is what it looks like when you are done with the dash. You saw that it is simply a matter of getting the wires figured out, shaping a harness for all of them and then installing male and female connectors so it is a quick in-and-out dash for easy maintenance. Once you decide where you are going to mount the tachometer, drill a hole for the wires and add a grommet to protect them.

26. Since the tachometer is mounted on top of the dash, mounting and wiring it is something you must do lying on your back under the dash. The painless kit has the tachometer signal wire and a tach power wire as well as light wires. It is simple to interface the call-outs on the Stewart Warner with the wires from the Painless Performance kit.

Chapter 12

Installing a High-End Audio/Video Deck

On this street rod, the radio is located in a custom console running from the inner firewall to the armrest between the seats. The location is fine for adjusting it when parked at a show or cruise event. Once the monster engine in this vehicle comes to life it is next to impossible to hear anything anyway so in traffic or on the road station changing isn't an issue.

This chapter deals with the installation of an aftermarket audio/video deck. We were lucky because there is an excellent custom auto sound store in our neighborhood. They allowed us to observe custom work on a Mini.

Installing a high-performance aftermarket audio system in a hot rod or a modern car is easier than you might think. On modern cars the hard part is getting the trim panels off so you can get at the factory unit. On older cars it is usually just a matter of getting on the floor and looking up so you can remove the rear mounting bolt(s), ground and the antenna wire.

On a hot rod, the system can be located just about anywhere. It might be in a console between the two bucket seats, under the seat on a roadster, in the dash, or even overhead in a special console. On a muscle car or cars that are more stock the unit will be in the dash. This makes getting to it, removing trim and other things, more difficult.

The Mini Cooper that Sound FX was going to work on already comes equipped with a good sound system, but the customer wanted great. Sound FX installed a Pioneer DEH-6,000UB. This deck includes an AM/FM radio, DVD player, and an iPod connection in the console so you can play your music through the system. It has a model-specific connector that also charges the iPod while it is connected. The Mini had already been modified with an external high-output amplifier and fantastic speakers by Sound FX. The audio unit was the last item to make the upgrade to the installation complete.

The complete installation took Sound FX about one hour, but for most of us, it will take about 3 hours. Although the step-by-step is demonstrated on the Mini, the process is similar for upgrading any OEM audio system. On most vehicles, there is a factory frame or bracket to hold the radio in place. In the case of muscle cars or classics, the radio is usually held in place by the control knobs and a mounting bracket in the rear. Under the dash you may have A/C and heater ducting blocking access to the audio unit mounts or any number of other components for other systems in the vehicle.

Once again, the hardest part of this job was getting to the radio. Knowing which panels cover what is the key. I would not have looked under the mirror switch panel for another two bolts, as you'll see in the how-to photos. If you plan on making this kind of a project a reality, go online and get as much info as possible for your make, model and year vehicle. Just take your time and be patient. Always look for the simple answer before you get into anything complicated.

1. The Mini Cooper already had a high-output amplifier and high-performance speakers. But the audio unit was stock and the owner wanted to complete his custom system upgrade. The choice of radio should always be based upon features and benefits, even though the look is also important to most owners.

2. From the outside the Mini may look very small but inside there is actually a lot of room. That is because of good packaging, but it also makes getting at things more complex. The key to any installation is knowing what pieces come off and how. To my way of looking at things, that is the reason to pay the pros to do special installations.

3. The interior of the Mini is surprisingly spacious in the front, but I wouldn't ride far in the back. The car is built for performance as the tachometer being in front of the driver and the speedometer being in the middle of the car demonstrates. The audio unit is suspended under the center of the dash with two supports going down to the transmission hump. This photo shows three layers of component controls; audio unit at the top, heat and A/C controls and a switch panel with five of the locations filled with simple and racing-type toggle switches.

4. We are installing a Pioneer DEH-P6000UB in-car entertainment system, USB cable extender for hooking up an iPod, OEM to Pioneer adapter, remote control, power cable adapter, antenna adapter, and odds and ends that complete the installation, such as custom faceplate adapters.

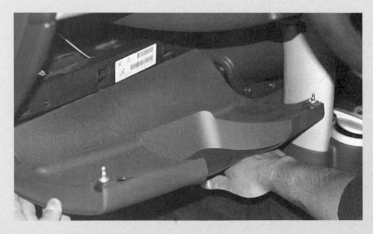

5. Start by removing the knee protector panel under the dash in front of the driver's chair. It snaps out at the top and then slides out of the bottom. Set it aside as it goes back in last.

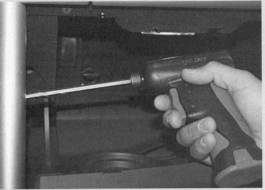

6. Next remove the trim pieces by taking out the screws in the bottom of the cup holders. Note the trick little electric screwdriver. It is made by Snap-on and although it is powered by AA batteries, it delivers enough torque for any small job like this. It really is a "must" tool for your toolbox.

7. Next, remove the two screws on the outside of the vertical supports. They attach the supports to the suspended layers of the center control unit.

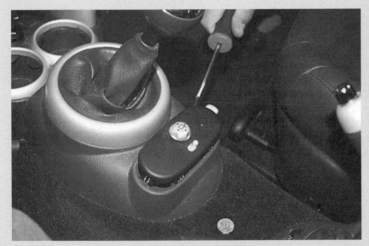

8. The last screws are located under the outside mirror control. Pry it up and it will pop out easily. Remember, no muscle moves. There is always a lot of pocket change on the floor of the car, so since you will be doing this on your own car, clean it up and put it in the piggy bank.

9. You can now pull the vertical supports towards the back of the car. Just pull it gently and it will come loose.

10. Once they are pulled back, slide them down and angle them to go along both sides of the shifter tunnel hump. Without too much effort you can now slide the cup holder assembly and shifter boot back a couple of inches. This gives you access to the mounting screws for the audio unit and all the control layers.

11. The mounting screws are Torx heads so change the bit on your screw driver or ratchet and extension. If you haven't purchased a set of Torx tools, you really will need them if you are working on anything newer than 1990. There were two screws on either side and they were quickly removed.

12. Now grab the faceplate of the audio unit mount assembly and slide it out of the mounting chassis. Just pull it out until the back of the unit is accessible. You might need to bend the chassis adapter locks down to remove the audio unit, but a few wiggles generally do the job instead.

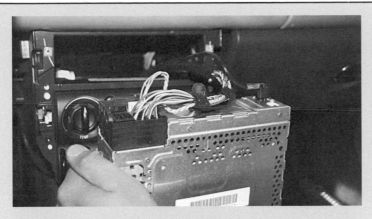

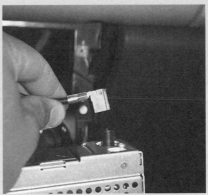

14. To get the antenna off you must remove the restraining clip. It snaps off and then you pull the connector out of the audio unit.

13. On the back of the unit you will find a master connector (in this case with a locking lever), and an antenna connector with a restraining clip. Take both of these off and you can take it to the work bench.

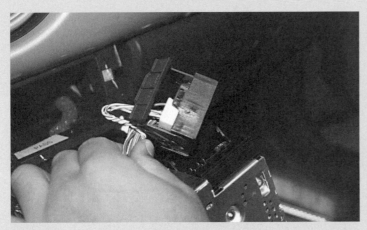

15. Lift up the master connector lock lever and it will pop up and you can pull the connector off of the audio unit.

16. The back of the new audio unit deck is different from the stock one. It has two types of connector choices for the speakers, amplifier and audio in.

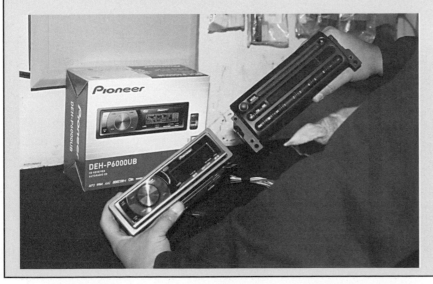

17. Looking at the stock and new decks, you can see that the new one has the CD slot hidden. It also seems to be missing control buttons for AM/FM audio unit and a few other functions. These controls are on the remote control not on the face of the Pioneer deck.

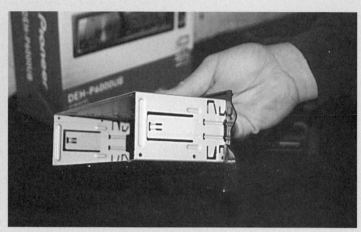

18. The control center chassis adapter for the unit needs to be removed from the old audio unit and installed on the new one. The pry tool used in this photo is really easier to use than a bent screwdriver. If you are going to do more than a couple of installations, get it.

19. This photo shows how the chassis adapter works. It is an essential component for the audio unit install. The chassis adapter locks were pushed down during removal and need to be pried up gently before installation.

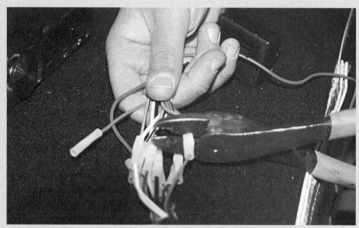

20. To be able to adapt the Pioneer audio system connector to the factory one, an intermediate adapter is required. You cut the connector off the old audio system and splice each wire into the correct location. The instructions included in this kit make it very easy.

21. This faceplate included in the Pioneer kit needs to be snapped into place on the chassis adapter and audio unit. It makes the system fit just like the stock unit. Note the screw holes are the same.

22. On the left is the USB adapter for the iPod to connect with. In his hand Javier holds the antenna adapter. On the right is the antenna connector. The antenna will not be connected until the unit is ready to be tested in the vehicle.

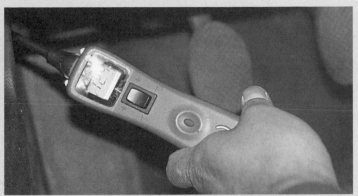

23. This tool will signal when a cable/wire has voltage and the reading will let the technician know how much is available. Once he finds voltage he will turn off the key to make sure it is "keyed voltage."

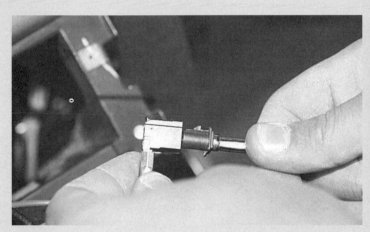

24. This is the other end of the antenna adapter being connected with the factory antenna. A simple task once the right connectors are together; without the adapter for the Pioneer to Mini antenna, it would drive you crazy.

25. Connecting the new locking lever master connector to the adapter connector is a simple snap-in job. The blue stuff is adhesive that locks the two together.

26. The locking lever master connector is now ready to install into the old master connector. Once it is in you are almost ready to test the unit.

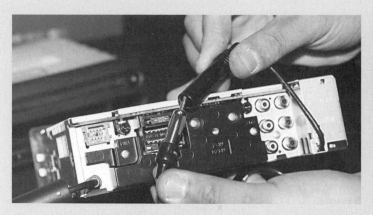

27. The antenna connector is snapped in place and it is ready for testing.

28. When testing make sure you check every setting of the unit. That includes AM/FM, Seek, Scan, Volume, On/Off, Base/Treble, Equalizer, Fader, etc.

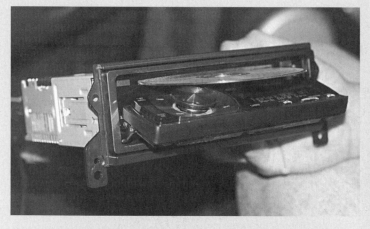

29. After testing everything on the face of the deck, hit the switch and the faceplate will open so you can slip in a CD. This photo gives you a clear look at all the features, including a removable faceplate.

30. An extension for the USB cable is attached to the short wire on the back of the audio unit. It is taped to keep it together under the most severe conditions. The other end was run into the dash before the audio unit was set in place.

31. Tape all loose wires together to keep the clutter under and behind the dash to a minimum before you slip the unit into the dash.

32. Now is the time to slip the new Pioneer deck into the chassis of the suspended control center. It just pops in.

33. There are four Torx screws to reinstall and it requires nothing else to hold it in place. Reverse the removal process for the installation. Get the upright supports back in place, move the cup holder back forward, install the screws under the mirror switch panel and the rest.

34. Once the interior is cleaned up and vacuumed out, you are done. Although this was done on a two-year-old Mini, it is just about the same for all modern cars.

Chapter 13
Trailer Wiring

This bike trailer was in poor shape, and although still useful, it needed a complete overhaul. We tackled the wiring, but it is eventually going to need some spot welding, paint and tires if it is to be remain safe enough to use for any length of time.

In this chapter we are going to talk about trailers and how to wire them from scratch. Although trailers seldom create debates in bench racing sessions, they are an important factor for most hot rodders—after all, how else are you going to get your show car or race car to the event? Regardless of the actual cost, you have a lot of sweat and money invested in your car. So, since you will have to use a trailer to take it somewhere sometime, you need to know how to make sure that trailer is safe for towing.

Trailer maintenance should be performed as regularly as the vehicle you use to tow it. At the least, you need to check the integrity of the mechanical and electrical components before you load it, hook up and take off.

Generally I recommend packing all wheel bearings before you use a trailer the first time. If it has been used to haul a boat, you will find that the wheel bearings will be packed with a bearing buddy. This is a type of pressurized grease system that keeps water out of the bearings when you are taking the boat off of the trailer. They are fine for towing on dry land; just make sure you grease them with high-pressure grease.

Regardless of the type used, you should periodically remove the wheels and repack the bearings, one by one. I will not get into the methods of repacking wheel bearings, but if you do not know how, take it to a reputable tire shop.

Electrical Inspection

Verify that the wires, grounds and the connector that ties the trailer into the electrical system of the tow vehicle are in good condition, and repair anything that looks worn or frayed. Look for cracks, chafed tubing and poor grounds.

After repairing anything damaged on the trailer it is time to test the system independent of the tow vehicle. I use a low amperage power supply and a power test lead attached to the power side after grounding the tester to the trailer.

You must decide how many wires and what you really need it to have, 4 or 6 wires. You can use the standard 4-wire connector that includes turn signals and lights to handle the needs of your trailer. Although some of my friends have wired their trailers using the tow ball as the ground connector for the trailer, I do not trust it to be a good ground. It may have a lot of rust in the socket and you will surely grease the ball before connecting the trailer and ball together. Aftermarket kits include a real ground wire from the trailer to the tow vehicle. They know the right way.

In the following pages is a step-by-step process of how I rewired a small motorcycle trailer that had been made using the chassis of an old pop-up camper. It is in rough overall shape to say the least, but it works and is used on a regular basis. Now that it is rewired it wouldn't be a stretch of the truth to say the most beautiful part of the trailer is the wiring.

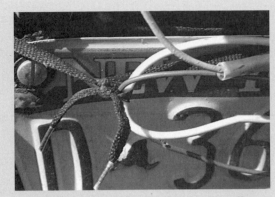

1. When I had the owner drop the trailer off at my house so I could rewire it, it was a shock to see the condition it was in. How long do you think this shoelace has been holding up the wiring? And, it was not the most serious issue I found.

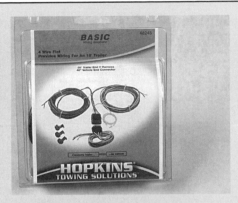

2. You will find a kit like this at your local autoparts store and at Wal-Mart in the automotive section. I paid $14.94 for the kit at Advance Autoparts and it included everything needed for the installation. They all have everything you need for light trailers and if you need something bigger, there are specialty shops that have these items. U-Haul is also a quality supplier and has just about anything you might need.

3. This kit includes 20 feet of five-wire harness for the trailer, about 6 feet for the tow vehicle. The instructions are good and straightforward. But, they do not prepare you for the mess I found on this trailer.

4. Most trailers are made from rectangular steel tubing for the frame. It is easy to run wires through them to keep them protected. This trailer already had the main wiring harness in the frame, making it easy to run the harness from the front to the back without going outside the frame. To get the new wires to the back of the trailer we first cut off the front part of the old harness going to the trailer/tow vehicle connector.

5. We found these four crimp splices in the harness up front and that is really strange. The only splices needed are from the harness to the taillight. Splices up front indicate problems.

6. This view shows the wires going into the trailer frame at the front and the four splices in front of that. That says "someone had run over and/or ruined the tow connector to the tow vehicle."

7. We had saved some of the old wiring in front as pigtails so we could splice the new wiring in and simply pull the old wiring out from the back, letting the new wiring be pulled into place with the old. It is much easier to do it this way and although you can use an electrician's wiring snake, the old stuff is easy. Note that we only used one splice for two wires and tapered the other wire so it would pull through easily.

8. Once we had the new wire at the rear of the trailer, we pulled it free of the frame through the stock holes. (There are two at the rear on each side for a total of four in the back and two closer to the front.)

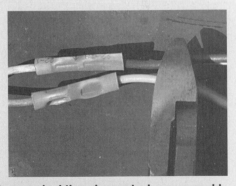

9. Once we had the wires out where we could see them we simply cut the old wires at the bottom of the crimp splice and threw the old ones away. Although this is a small utility trailer every step in the rewire job is the same as you would need for a 28-foot trailer. The only differences would be electric brakes, extra running lights and interior lights. I actually think it is easier on a bigger trailer because things are not so tight and they are constructed better.

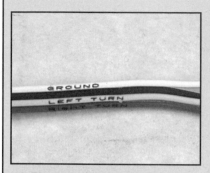

10. The connector and wires that go to the tow vehicle have their function printed right on them. Ground (white), Left Turn (Yellow), Right Turn (Blue) and Power (Brown). This is about a 10- to 15-minute install on the tow vehicle.

12. In this photo you can see the wires going to the tail lamp on the driver's side. If you look at the bottom right you will see the wires having to pass over sharp crusty old steel. Not a good situation and sooner or later it will rub through the insulation and a short circuit will occur.

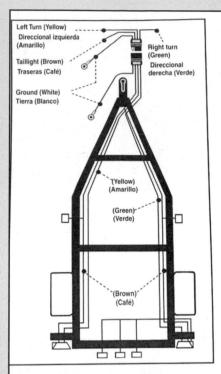

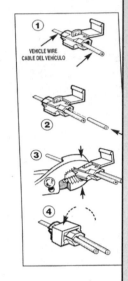

11. At this point you need to determine which wires go on the driver's side of the trailer and which ones go across to the passenger's side. There is a schematic with the kit showing the trailer frame and a guide telling you which is which. Follow it and you will have no problems. It doesn't matter which wire is ground and which is power in the lamp as long as you match colors to make trouble-shooting easier for the next person trying to fix it. Courtesy Hopkins Manufacturing.

13. We started by cutting the wires to the tail lamp on the driver's side leaving as big a pigtail to work with as possible. We found that the wires had been spliced by twisting them together and taping them. They should have been spliced by solder or by a crimp connector. In both cases shrink tubing should have been installed over each splice and heated to seal it from the elements. We spliced in the yellow wire and one of the power wires (brown to the driver's side, yellow is left turn signal).

14. This photo sequence shows the correct way to solder wires together and seal them. We staggered the wires a bit so the splices wouldn't be right on top of each other when we bundled everything up into a harness. Why am I wiring a green wire to the yellow from the new kit? Because both of the tail lamps (driver's side and passenger side) were brown and green. I knew that the yellow is for the driver's side left-turn signal so I wired the green and the yellow together. On the other side there is another lamp with brown and green. In that case wire the green to green and brown to brown.

15. These two photos show what the splices look like when they were sealed inside shrink tubing. They come out of the lamp assembly in the back and the stagger we used in the splices will make sure they bundle well in the protective split tubing as they are made into a simple harness.

16. I was surprised at all the sharp edges that threatened to damage the wiring throughout the chassis and from one side to the other. To fix this we decided to run protective split tubing over all the wire in the trailer. Here you can see how we prepared the wires to go from the driver's side to the passenger side under the chassis.

17. The split tubing goes over the wires and splices forming a strong protective barrier for the wires so they do not rub anything sharp. Although I am taping the tubing to keep the wires from getting out, this tubing will not keep wires waterproof. That is done with the shrink tubing over all splices and other connections. I also used Permatex Blue, RTV Sealer/Gasket Maker silicon to seal each of the lamp bodies in the back where the wires enter.

18. Since all four wires came through the frame channel on the driver's side, I needed to run one power wire (brown) and the correct wire (green)for the right-turn signals to the passenger side of the trailer. They were run through a "V" notch formed by one of the support runners from the back of the trailer. This was a good place to run the wires, but not without protection for them. The originals were frayed and shorting out.

19. Here you can see the wires for each side coming out of the access hole in the frame channel in the rear. The one on the left is from the driver's side left-turn signal and running light, and the one on the right does the same for the passenger side. They are already inside split tubing and mounted securely to the floor of the trailer with an insulated clamp.

20. Now the wiring is safe going through the "V" notch all the way to the passenger side. The right way to do things is just as easy as the wrong way. The split tubing used on this vehicle was less than $10 dollars.

21. We cut out these splices and did the same kind of soldered splices on the passenger side and also sealed the back of the lamp housing.

22. When we were done with the soldering and shrink tubing installation, we ran the rest of the split tubing up to protect the wires on this side.

23. The ground wire from the old wires looked fine, but I had to replace it anyway. Once it was removed you could see that there was little contact between the ground wire and the frame. Why? Rust under the eye terminal. The fix was easy, some sandpaper on a die grinder cleaned it right up. I also ran a tap through the threads of the attachment point. We added shrink tubing over the eye terminal and attached it right back on the existing location for the ground.

24. At this point we were done with the project. Instead of checking it out with a non-automotive power source, I connected it to my pickup for testing.

25. I hooked the trailer up to my vehicle to test all of the lights, and everything worked just fine. The new wiring is really the best looking part of this trailer and it will reassure the owner he has safe, legal lighting every time he hooks up to it.

Chapter 14

Electrical System Troubleshooting Tips

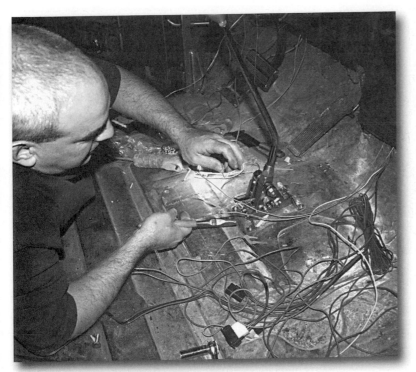

Lying on a bare floor as you wire a vehicle from scratch is far easier than finding problems after everything is installed and covered with carpet, noise and heat padding, kick panels, door panels, consoles and seats. Because of that you need to have a diagnostic strategy, a plan that will allow you to eliminate problems from the easiest to the hardest possible problems so you can find and repair even open circuits in a harness. We have outlined a three-step process that will help you solve problems like a professional.

BASIC ELECTRICAL DIAGNOSTICS

Sophisticated automotive electrical systems, as well as those on hot rods, street rods, muscle cars and race cars, can increasingly be the cause of dead batteries, intermittent warning lights and general electrical complaints. When addressing these issues, if you stay focused on the basics you will save yourself time, money and frustration. The three most important automotive electrical diagnostic basics are:
• Connection condition
• Battery condition
• Alternator condition

On today's computer-controlled vehicles, you can sit in the driver's seat and do your first step in chasing problems by connecting to the OBD II port with a scanner. Depending upon your equipment, you may just get codes, or if it is more sophisticated, it will give you actual problems to sort through. If your cable is long enough you can also do this outside the driver's compartment looking at the engine and other components as well as the problems the scanner finds.

Connection Condition

In order of importance, connection condition (at least to the battery and major ground and power wires) really does come first. Without a good solid connection to the battery

You can do a lot of diagnostic work with a scanner before picking up any tools and tearing things apart. This equipment will give you problem information, not just codes, as well as point you in the right direction to solve the problems.

When it comes to starting with the basics, this is where it starts. Believe it or not, this was how I found the battery ground cable on a professionally prepared and maintained race car. The driver and owner had been complaining about poor performance that went away sometimes and returned on its own without any warning. In this magnified photo you can see that the ground terminal is cracked in half and was at least one-half of the cause of the problems. This was caught before any testing was done, just by performing a visual inspection. Look first, test second.

That broken terminal end was replaced by this end that gave us the ability to take one wire from the chassis ground common stud inside the car to the battery ground wire.

and ground, the vehicle may not start, may not charge, or it may generate unwanted electrical gremlins. Clean tight connections from the battery to the chassis grounds, and to the alternator and to the starter motor will eliminate even the most mysterious gremlins and most minor electrical problems. Do not forget a visual inspection before you grab your tools.

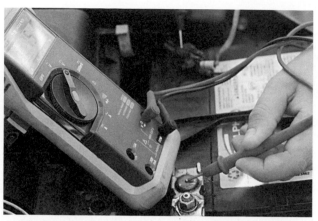

A voltage drop test will let you know if the battery has enough current and amperage to start a vehicle. It will tell you how much current is available for a component after surface/flash voltage is drained off of the circuit, and a host of other component tests. First test the circuit for voltage before doing anything else. Note the voltage available on a slip of paper. Then turn on lights and accessories for 1 minute to drain off any flash charge and then test it again. The difference is the voltage drop.

The condition of all electrical connections, and the circuits they connect to, can be easily tested using the voltage drop test. You can find bad relays, bad fuses, poor connections and faulty ignition switches within minutes (as opposed to hours or days). This may be the single most important test for you to integrate into your diagnostic approach and strategy when searching for the source of any intermittent electrical complaint.

Battery Condition

Battery condition is paramount on today's vehicles because the on-board computers will be the first to react to a problem within the battery or charging system. Computers are reliant on continuous, uninterrupted current flow between the positive and negative terminals of the battery. But even with a carburetor vehicle, instead of a computer, there are untold problems that you will encounter with bad connections and a bad battery.

One bad plate in the battery can cause serious electrical issues after you start the engine. You may be thinking that if the battery was bad the starter motor wouldn't crank the engine, but that is not true. Check the condition of the battery to ensure it is not the cause of the fault. The voltage drop test is the key to testing a battery.

Alternator Condition

After a battery has been determined to be in good condition, you will need to determine the condition of the alternator. With the number of electronic circuits in custom vehicles, the alternator output

Regardless of the type, every automotive 12-volt battery will have 6 cells that will store 2 volts or more. More importantly, a quality battery will provide a long service life and handle the requirements of the vehicle over the period of 3 to 5 years in normal operation.

needed on some cars may exceed 200 amps if everything is on and at full blast. With that much output required, the main power lead to the alternator may need to be fuse protected. Before replacing an alternator because of "no-output," double check the manufacturer's information regarding their test procedures. You may also discover that a fuse, circuit breaker, fusible link or specially designed cable is the cause of the no-output problem.

One other note: some late-model commercial vehicles could have two alternators and/or two batteries that are networked and diagnosing them will be different. It is not uncommon to replace a factory 40-amp alternator with a 100-amp aftermarket unit. Even on race cars, this may be the case, because high-tech ignitions, lights and other systems can really use up the juice in a hurry.

Checking the Alternator—There is a quick check you can do before taking the time to voltage drop the alternator system. It is very simple and although not definitive, it does give you quick information. Use jumper cables to start the vehicle. Once it is running and warmed up, remove them. If it keeps running, the alternator is probably fine and you should take the battery to your auto parts shop to have them test it under load before you replace it. If it turns out to be good, make the following checks to the alternator.

The only other thing that could be wrong with the alternator, if it could keep the engine running without the battery being there, is the diode/triad, or rectifier. To properly test the alternator you must start with a fully charged battery. With the engine stopped, verify that there is a no-load voltage reading of between 13.8 and 15.3 volts with your digital multimeter reading across the two battery terminals.

Next, start the vehicle and add load to the alternator for 1 minute by turning on as many

MEASURING BATTERY VOLTAGE DROP

If your battery becomes discharged it will be unable to supply enough voltage/current to the starter. In that case the starter will not crank (turn over enough to start the engine). Neither will it power the computer or a myriad of other accessories that may be switched on while trying to start the engine. Verify that the connections are clean and secure then test the voltage in the battery and write it down for reference later. Then, with the engine off, bleed the surface charge off the battery by turning on the headlights and other accessories for a minute or two. Now, with the lights and accessories off, set your digital multimeter to the DC voltage function 20-volt range and measure voltage across the terminals by connecting the red wire to the positive side and the black wire to the negative side. Once again, the difference between before removing the flash charge and after removing it is your voltage drop. The readings at 80 degrees Fahrenheit of ambient temperature after removing the flash charge should be as follows:

12.6–12.75 volts =	100%
>12.45 volts =	75%
>12.3 volts =	50%
>12.15 volts =	25%
<12.15 volts =	Discharged

Note: All of these readings show more than 12 volts while the usability of the battery is significantly different at each step. Just because you have 12 volts, you cannot count on the battery being operable. At this point it is time to charge the battery, if it is below 100%, before doing any more testing.

This stock 12-volt alternator is just a 40-amp unit on a small import car. It is tightly packaged in the engine bay and is driven by the serpentine belt at the front of the engine, as is the water pump, power steering and A/C. Make sure the alternator is turning when the engine is running. If not, it could be a frozen alternator bearing, a frozen water pump or that the belt is missing.

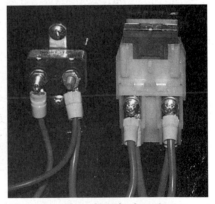

All automotive electrical systems usually contain a high-amperage fuse that goes between the main power from the starter and the rest of the electrical system. This photo shows both a 75-amp fuse (right) and an electrical circuit breaker (left) that will go "open" when overloaded so it can protect the circuit. (It will return to the closed position when the circuit is operating and it stays within its heat parameters.) Make sure both of these are keeping the circuit closed, before looking elsewhere.

This is where the high-amperage fuse resides in a stock Acura hatchback. The fuse directly below the fingertip is a 100-amp fuse that provides the main circuit protection from the battery/alternator.

This photo shows the wiring on a GM three-wire alternator. The three wires are the BAT wire (taking the voltage from the alternator and returning it to a power buss or to the battery), the exciter wire (the small dark wire that takes a little DC current into the alternator "field-in" windings to make it active when it is turning) and the white wire (field-out). GM also has a single-wire model that makes these connections internally.

Just checking ohms from one end to the other end of a wire will let you know if it is open (bad) or closed (good). When checking spark plug wires you need to check the differences between each wire. If they are all close they are good; if not, one or more are bad. The higher the ohms, the more resistance to current flow and that is never good.

multimeter ground probe to a good ground and the power lead to the output terminal of the alternator (usually marked BAT). If there is more than 0.50 volts AC the alternator probably has a bad diode. It is easier to replace the entire unit with a good rebuilt than repairing the old one yourself.

Checking for Circuit Integrity

Every electrical system demands secure full-contact connections to pass current from one point to another. Keeping this basic thought in mind will help you diagnose and locate problems. Electrical systems on vehicles are subject to a lot of vibration, corrosion, impacts from road debris and, to be candid, they operate in a harsh environment. Therefore troubleshooting the integrity of each system is critical when there are problems.

The starting point is a test with a digital or analog multimeter to see if the electrical device is receiving power. Turn the vehicle ignition on, then set the meter to DC current (20 volts if it has individual settings) and find a good ground near the component. Connect the black ground probe to the ground and the red probe to the power connection. You should be getting at least 12 volts to that power lead. If not, you need to first check for a blown fuse and if one is blown, replace it. You also need to look for what might have caused it to blow. If the component/device is not working, even with power to it, you must now check for good ground(s). Without a good ground all the power in the world will not make it work. You can use your digital multimeter to make a circuit from the device ground to a good ground nearby. Place your red probe to the ground wire connection and then use the black probe to reach a real ground. If the component/device starts working you know it is the ground. If it still doesn't work the component must be faulty and you will need to replace it. (Remember, this goes for any component, from starter to radio.)

Take the red lead and connect it to the ground wire out of the component (unless it is self-grounding) and then take the black lead to a good ground. If the unit starts working you know the ground is bad. If the unit is self-grounding and you place the red lead to the component and the black to a good ground and it starts working you know the mounting screws/bolts are corroded and not making ground. Trace the power lead back to its source. As you do so, check for loose connectors, broken wires and corrosion at every junction back to the power source. Fix any problem(s) until you get power to the component. Checking for power at each junction is also a good way to test for open

accessories to the system as possible (headlights, sound and video systems, driving lights, navigation system, heater, A/C, rear window heater, etc.). Raise the idle to at least 2000 rpm (no more than 2500) and check voltage output at the alternator output terminal (it usually will say "BAT"). It should be 12.6 volts with all these systems and accessories on. If it is at that level, the alternator is doing its job. Now you must look elsewhere. (The "diode triad" that converts AC to DC could still be bad.)

With the loads still applied and the rpm at least 2,000, turn your digital multimeter to AC voltage but no more than 2 volts are needed. Most digital meters will simply have a "V" with a "wave" line over it to check your AC Ripple Voltage. How alternators create DC current isn't important, but there should be some AC current present. Connect the digital circuits. Remember, if you do not have power at the component, before you waste any more time make sure the accessory switch is on.

If you find a bad ground, fix it and the problem will probably go away. If you find an open circuit, it is simple to just replace that segment of wire with the same size and new connectors. You can always go into the harness and fix the original wire, but why. Replace the wire between junctions and tape it into the original harness.

Checking Voltage Along the Way

To check wires and relay sensors, the first step is to have current flowing through the circuit, connection or component. Once that is set up, take the multimeter and set it to DC voltage in the 2–3 volt range. Now, connect the red lead to the side closest to the battery positive (+) terminal and the black lead to the side closest to the negative (–) terminal. What you will get is an mV (milliVolt) reading that should be within the ranges shown below. Anything higher indicates abnormally high current draw and a faulty component, which you will need to change.

- Switch/Relay: 300mV/(0.3V)
- Wire or Cable: 200mV/(0.2V)
- Ground: 100mV/(0.1V)
- Sensor Connections: 0mV to 50mV/(0V to 0.05V)
- Connectors: 0mV/(0V)
- Wiring: 0mV/(0V)

In many cases high readings on connections can be caused by corrosion or dirt and sometimes can be repaired by cleaning them. This is especially true with ground connections. Sometimes just twisting them or disconnecting and connecting them again can be all that is needed to eliminate the problem. But, switches or wiring that show high voltage drop readings should be changed.

Checking Thermal Switches

If your computer doesn't receive engine operating temperature information because of a faulty thermal switch, it will start, but it will stay on the rich setting needed for cold starting. This will cause poor operation and a poor idle. Basically a thermal switch is a temperature-sending unit with a circuit that opens (turns off) when it is cold and closes (turns on) when a specific temperature is reached. It can also be closed cold and open hot, but they are unusual. The sensor will be mounted in the radiator or in the intake manifold.

Either one should show different readings

Here a professional technician is working on cleaning up some electrical paths. Notice he has gloves on; these will protect his hands, keep them clean and they also make him non-conductive.

between hot and cold. It may look just like an electrical temperature gauge except it sends signals to the computer or any other device that operates differently between cold and hot temperatures. Check to see if there is one on your hot rod. To check it, you need to set your multimeter to continuity or ohms reading and see if there is any current flow between the two connections when the engine is cold. If it is showing open when cold proceed to the next step.

Now you need to test when the engine is hot. You are looking for the opposite reading to the first one. Warm up the vehicle and when it is at full operating temperature, check the reading again. If there had been voltage flow when cold, it should now be open with no voltage flow. If not, it is defective. If it had been open, with no voltage flow cold it should now show full flow. If it is still open and there is still no voltage flow it is defective. In either case if the thermal switch is not working properly, if it doesn't show different readings between cold and hot, replace it with a new one. You should find an operating temperature stamped on it and it will also have a code regarding how it works.

Common Electrical Path Problems

Troubleshooting is generally simple with electrical paths, yet sometimes it can be confusing. Just use a multimeter, test light and a schematic of the circuit and take your time.

Short Circuits—When working on a vehicle with electrical problems, you will likely encounter short circuits, which means that a hot or live wire has been grounded somewhere before it reaches the component it is supposed to power. This might be because the insulation gets damaged and the wire itself touches a ground or it can be from other causes. (Insulation is designed to keep the electrons

This trailer wire has the insulation ripped and every time it bounces against the frame of the trailer it shorts out. This results in a burned fuse and no lights. (That can result in a ticket.)

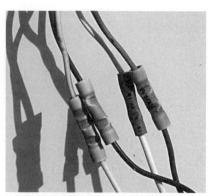

One of these 4-crimp connectors was causing an open circuit. It takes time and patience to work with a multimeter and check out a circuit wire by wire.

Engine electrical operations can be influenced by how the cylinder head is bolted in place. If you use a lot of sealant and pipe sealer on the head bolts you can lose conductivity, causing the plugs to not fire properly.

inside the wire running in a specific direction.) For example, if a wire carrying 12 volts from the light switch was grounded to the frame, before the light bulb, you would have a short circuit. That would cause sparks and would hopefully burn the fuse for that circuit (if everything goes as designed) before it burned up the wiring. If that same circuit went to ground after the light bulb there would be no problem, because the system is designed to go to ground after the bulb.

Open Circuits—You will also run into open circuits, where a circuit is interrupted before it reaches the component. This is often because of a loose connector wire, a break inside a live wire (where the insulation doesn't open up) before the component, a loss of a ground after the component or a loss of electricity to the circuit. These are not dangerous to the components because no power is involved in the problem. But, this type of problem is tougher to diagnose, so if you follow the circuit with a multimeter you will find the problem quickly in most cases.

Intermittent Operation—The most difficult type of problem to solve is intermittent operation, a situation where the component operates occasionally. The component, like a brake light, might be fine when you test the circuit in the garage, but then it won't work when driving down the road. The only way to find these problems is to identify which circuit and wires are involved and then pull them and move them and see if the circuit goes open. If it does, you will have to identify which wire it is before you can repair it. Just remember, 85% of all electrical problems are caused by bad grounds. This is especially true when it comes to open circuit and intermittent problems. You can use your multimeter and/or test light to

check where the circuit has electricity and where it doesn't. I like to start at the back of the circuit— where it goes to ground—and then work forward. I test the ground first, always.

Conductivity—When cylinder heads are bolted on an engine it is important that there is good conductivity between the head and the block. Spark plugs have only one wire going into them, so they must be self-grounding to work. The spark plug will fire because it is grounded to the head, but without good connectivity between the head and block, you will have an open circuit. There is another element to this circuit also, the engine block must also be grounded to the frame or directly to the negative side of a battery to work. If this ground system isn't complete, the plugs cannot fire and the engine will not start. If the grounds are just poor it might run, but poorly and without any power.

A bad ground situation can exist inside or outside the vehicle, on a trailer or between components. Most circuits are grounded to the body of the vehicle (auto, truck or trailer) using sheet metal screws. These grounds are not sent back to the negative side of a battery, so the body must be grounded to the chassis and then to the battery or directly to the battery for everything to work. This is why it is so important to check the grounds of a circuit first in the problem-solving process. These principals are in effect in any vehicle and any circuit and they are the most common of all electrical path problems.

There are also many other electrical paths that can become problems. An example is a simple switch that is placed in one position for off and another for on. These types of switches can fail internally and that could result in an always-on situation or an always-off problem. In both cases, unless you can feel the switch operating incorrectly, all it takes to isolate the problem is a multimeter or test light and testing it for current flow both off and on. Relays can become electrical path problems also, because they are nothing more than an electrical switch. Although a relay may only need a few amps to operate, that relay can control many times more amps because of the circuits connected to it. (Typically heavy amperage will be at the starter.)

Grounds

Grounds are critical to all circuits and are the major suspect when circuits do not operate correctly. About 85% of all dealer and repair shop complaints about poor performance and things not operating correctly are directly caused by faulty grounds. When bizarre things happen in systems or components you should look at the grounds first, as they can cause intermittent and open circuits as well as all kinds of trouble codes out of the computer interface. Most engine and radiator electrolysis damage can be reduced by grounding both, with ground wires that have a diode (one-way electrical check valve) installed to the radiator and heater core so current can flow out of components, but not back into them. Grounds are so critical to every circuit in a vehicle that you could say there is no such thing as too many good grounds. But in reality it isn't necessary to go crazy with them.

Once again let me stress, look to the grounds first if you have a problem. On vehicles that have the battery relocated to the rear, it is imperative the engine and battery both ground to the chassis and that nothing interferes with bundled grounds from circuits. They must make solid contact with the frame or unibody chassis, without paint or corrosion getting in the way. On full-frame vehicles, you must run a ground strap from the body to the frame to insure that everything from the interior of the body gets grounded properly to the chassis.

Most vehicle grounds are not sealed from the factory and, over time and miles, will lose their efficiency due to corrosion, miles of road grime or oily grease buildup. Your vehicle's performance is dependent on these critical grounds and as your vehicle ages, it can lose performance, fuel economy or suffer from drivability problems.

A good ground system can improve the efficiency of the vehicle's onboard computer by providing a direct ground path for sensors which relay messages to it, allowing for a much clearer signal for changing driving conditions. A ground wire system with special one-way ground wires will let the

This trailer electrical ground had worked properly for years, but all of a sudden the lights would not burn. The cause was corrosion under the eye-ring connector between it and the trailer frame. The same can happen inside a vehicle.

current flow out of the key components and not back into them. It can also stop computer spikes as current from other systems can no longer get back into the circuit through the grounds. These one-way ground wires to the heater core and radiator also help to prevent electrolytic metal corrosion, a destructive force that eats away at vital cooling system components.

Electrolysis/Electrolytic Corrosion

Electrolysis has been a problem for cooling systems since the first engine was built and dissimilar metals were used.

This process is when the coolant actually becomes and acts like the electrolyte in a battery. This is where two dissimilar metals (generally aluminum and iron) form the positive and negative poles inside the engine. (In other words, your cooling system will actually create electrical voltage, up to 2 volts and will become a small storage battery.) You can test your cooling system by taking a multimeter and sticking the red lead into the coolant and the black to ground. You will be surprised at what you find.

As happens to a battery's positive and negative plates, in this case, the aluminum inside the cooling system will start sulfating and that same white powder you find around the posts of your battery will start building up inside the cooling system. Since the softer metals get eaten up first, it is the aluminum in the heads, water pumps, intake manifolds, freeze/core plugs, radiators and heater cores that get damaged. These can become very

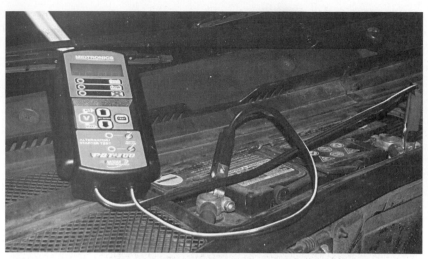

Taking advantage of new technology is always a good idea. Stores like Autozone, Advance Auto, Pep Boys and others will generally have this Midtronics tester or something similar. It will check your starter, alternator and battery (as well as their wiring) at the same time. If you want a quick answer and these stores are close, the test is usually free.

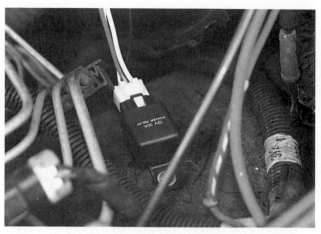

Look for things like this little black box. This is a typical automotive relay and they can be anywhere, especially if aftermarket equipment has been added. A quick voltage drop test will let you know if they are bad or good.

Always find the main power distribution/power buss of a vehicle before starting diagnostics. You will probably want to see if there is current available there before you look to individual circuits.

expensive to repair and if a simple ground kit can help stop the damage, it should become a priority. Getting the electrical flow out of the system to ground and not letting it back in can really save you money.

Take a look at the grounds on your vehicle and then take the time to clean and repair them. If you are wiring your vehicle for the first time, make sure you select ground locations that go to the chassis. If they are insulated from the chassis or the body, you will need to locate a better ground location and bundle several circuits to it. A common ground for several circuits will not cause a problem and you will also be able to diagnose problems quicker this

way. It also pays to check them from time to time (at least once a year) as preventive maintenance is easier than being stopped on the side of the road.

We rewired a race car that was driving the owner nuts because it would suffer intermittent electrical problems. It would almost quit running, then run great, then run rough. It turned out the ground cable terminal at the battery was cracked, the ground was poor, the problem major.

Other Troubleshooting Tips

Vehicle Will Not Start—If the engine turns over but won't start, your problem is either air, ignition or fuel delivery. Make sure all of these systems are functioning. I've seen a mechanic struggle with a vehicle that wouldn't start for an hour only to find there was a rag in the air inlet. Check the simple stuff first.

Engine Won't Turnover, Even When Jumped—First of all make sure your jumper cables are big enough. If in doubt use two pairs. If it still will not turn over it is probably the starter and/or the starter gear or ring on the flywheel. It could also be the positive cable coming from the battery to the starter, or one of the small wires in the circuit that comes, or goes, to the ignition key switch is damaged. (Wire damage is something that can usually be found by a visible check as it will be a large cable broken or it will be a connector or a small wire going to the starter solenoid that is broken off or even missing.)

Miscellaneous Tips—If the body of your hot rod is fiberglass or composite materials, you need to have a ground junction block in the vehicle and then a large ground wire going from it to the chassis.

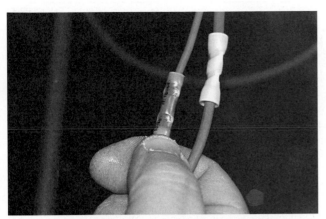

Do not be afraid to tug on wires to see if they are loose or broken under the insulation. If they come apart, you can always reconnect them and you will generally find that some of the problems are fixed also if you are chasing electrical gremlins.

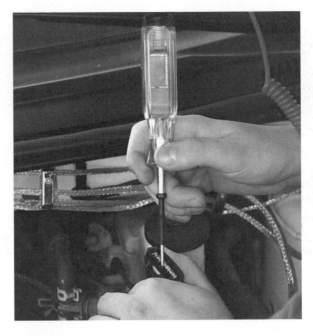

If you want to look for problems quickly, this test light with a voltmeter can take the place of the multimeter in many tests.

Body bolts on a steel body cannot be considered as viable grounds. You MUST run a large wire ground to the chassis from the body to make sure you have a good ground. The same goes for the engine. The engine must be grounded directly to the chassis. Motor mounts are made from good insulating materials and cannot be considered a viable ground. Just remember to take diagnosis of electrical problems step-by-step and keep it simple. Use deductive reasoning as it will help you get through most common problems.

Conclusion

We have covered all the basics of troubleshooting electrical problems. There are many components in a car and they are increasingly difficult to isolate and test. But your strategy must be consistent as we have explained. Look at the simple things before looking at the expensive and complicated things. Remember that grounds make up the majority of electrical problems so you should always check them first. Have fun and don't let fear be your response to electrical problems.

BASIC RULES

Basic Rule #1: Read the instructions first. Most don't adhere to this rule until "all else fails," but by then you will have likely wasted a lot of time. Read the instructions before you tackle a wiring install and you'll be happier in the long run. Besides, no one has to know!

Basic Rule #2: Expect to make mistakes, then learn from them. You will not always get it right on the first try. It is how you react to a mistake that matters, not what you did. So when you make a mistake, do not panic. Everything you will do to a car can be corrected, it's just a matter of time and money.

Basic Rule #3: Always run wire through a metal hole with some sort of grommet/insulator between the metal and the wire jacket. If you do not, in time it will chaff through and short out. There are all sorts of grommets available.

Basic Rule #4: Always take care when stripping insulation. NEVER cut into the cable. If when you strip the insulation you take a few strands of copper as well, do it over. You have just changed the resistance of the wire.

Basic Rule #5: Always size the wire to the job. Example, the alternator or generator will generally need a #10 gauge wire. The lighting is generally #14 or #12. You should use #12 for a heater motor.

Basic Rule #6: Always use a crimp connector or solder and add a heat shrink cover to join wires. Avoid electrical tape, which dries out in time and may come off, thus exposing your wiring. Solderless/crimped are common and when correctly applied are excellent. Wire nuts are only a last resort emergency/temporary fix. Keep a couple with your spare fuses as they can get you going on the road, but fix the wires properly when you get back to your garage.

Basic Rule #7: Always route your wiring away from heat sources like exhaust pipes or turning hardware like fans or drive shafts.

Basic Rule #8: Always secure your wiring or tie it up. It is best to harness any of your wiring with wire ties or tape to be sure they are secured and not easily caught on something. We use the Painless Wiring PowerBraid a lot, but there are many types available.

Basic Rule #9: Always keep your grounds as short as possible. It is far better to have a few more connections than a long ground that can build heat and introduce ground loops. It is common practice to run a ground wire from a high-demand device such as a fuel pump back to the battery. This wire will reduce the load to the pump or device and also reduce the possibility of pump failure.

Basic Rule #10: If at all possible, use stainless hardware to fasten all primary cables. Some galvanic or plated hardware can cause dissimilar corrosion and can sometimes carry more resistance than its stainless counter parts. On smaller secondary circuits, this is not as important.

Basic Rule #11: Never use THHN/THWN solid core wire in any automotive circuit. This wire classification is not rated for automotive use. Cars move and vibrate which chafes the insulating material and fatigues the wire until it breaks. The same with single-core house wire. It does not do well with vibration.

Basic Rule #12: A wire that is rated for A/C voltage may have a different rating when used in low voltage, D/C applications. Your house uses A/C, your car uses D/C. Make sure the wire is rated for the correct use.

Basic Rule #13: The best primary power cable is a type TEW or MTW (Machine Tool Wire) as it is rated for continuous use with higher amperage loads.

Basic Rule #14: OFC (oxygen-free copper) is by far the best cable for automotive use. However, it is expensive and requires a quality crimper for security. The most common application for this wire is high-end car stereo applications.

Basic Rule #15: If you are like me, you may choose to skip OFC and go with TXL wire. TXL (thermal cross-linked) wire is next best. TXL wire has twice the voltage rating of standard GPT (general purpose) wire. Most high-end harnesses are made with this wire. If you want to keep your ride for the long haul, you may want to pony up for a harness made with this wire. If you are just fixing a daily driver you do not plan on keeping, go with GPT, it is less costly.

Basic Rule #16: Most electromechanical devices are rated according to the MTBF (Mean Time Between Failures). You improve your odds of a long MTBF by keeping your electrical devices in a low-humidity, low-heat environment.

Basic Rule #17: Always use the right tool for the job. Not as obvious as it sounds.

Basic Rule #18: Soldering is a skill that must be practiced. You must make sure all joints are properly wetted by the solder. You must make sure that the insulation does not get burned. Remember to use an appropriate wattage soldering iron. Allow for the ambient temperature. You will find it very hard to get a good joint when the temperature is below zero or you are working outside with a cold wind blowing across the work area.

ABOUT THE AUTHOR

Matt Strong started working on cars and trucks with his grandfather when he was ten years old, and since then he has been involved in every facet of vehicle building and repair. He spent four years working on jet aircraft and visited Southeast Asia for two years. He worked as a line mechanic for a Chevrolet dealer, then seven years with AMC factory support in road racing and drag racing. Then he ended up back at GM Engineering in St. Louis when the Corvette was still being assembled there. At about the same time, he decided to start writing for car magazines, first for the industry publication *Hot Rod Industry News,* and then for other magazines in the original Petersen Publishing Group.

Matt worked in Mexico and South America for seven years at Ogilvy & Mather, a large international advertising agency. He did work for Uniroyal Tires, Volkswagen de Mexico and other automotive accounts using his extensive technical background to write ad copy and promotional materials. At the same time he began writing for enthusiast magazines in Latin America. Matt also became involved with Pro Rally driving in Mexico. He returned to the U.S. and went back into writing about cars, light trucks and even big diesel rigs for several magazines. He even spent two years teaching automotive technology at Richmond High School in Northern California.

He now lives just outside New York City. For more about him you can visit his website at www.matt-strong.com.